Root Cause Analysis

Improving Performance for Bottom-Line Results

Fourth Edition

Root Cause Analysis

Improving Performance for Bottom-Line Results

Fourth Edition

Robert J. Latino ▪ Kenneth C. Latino ▪ Mark A. Latino

CRC Press is an imprint of the
Taylor & Francis Group, an **informa** business

CRC Press
Taylor & Francis Group
6000 Broken Sound Parkway NW, Suite 300
Boca Raton, FL 33487-2742

Printed in the United States of America on acid-free paper
Version Date: 20110518

International Standard Book Number: 978-1-4398-5092-3 (Hardback)

Visit the Taylor & Francis Web site at
http://www.taylorandfrancis.com

and the CRC Press Web site at
http://www.crcpress.com

In Loving Memory of

Charles J. Latino

Known as the "Father of Manufacturing Reliability" for his pioneering efforts to break the paradigms of reaction and foster a culture of proaction. Charles' efforts in the 1960s paved the way for his enduring Reliability principles to be applied to any industry today. While he was known as the "Father of Manufacturing Reliability" to most, more importantly he was our loving father as well and we miss him dearly.

Contents

Preface

What is Root Cause Analysis (RCA)? It seems like such an easy question to answer, yet from novices to veterans and practitioners to providers, we cannot seem to agree (nor come to consensus) on an acceptable definition for the industry. Why? We will discuss our beliefs as to why it is so hard to get such consensus and why various providers are reluctant for that to happen.

Many who will read this text are seeking to learn the basics about what is involved with conducting an RCA. Many veterans will peruse this text seeking to see if they can find any pearls of conventional wisdom that they do not already know or to dispute and debate our philosophies. This creates a very broad spectrum of expectation that we will try to accommodate. However, in the end, success shall be defined by the demonstration of quantifiable results and not on adherence to the approach of favor.

We tried to write this text in a conversational style because we believe this is a format that most "rooticians" can relate to. Basically, we wrote like we were teaching a workshop.

Readers will find that much of our experience comes not only from the practicing of RCA in the field, but more so from our experiences with the over 10,000 analysts whom we have taught and mentored over the years. Additionally, we participate in many on-line discussion forums where we interact with beginners, veterans, and most providers for the betterment of the RCA field. We list these sources in this text in the hopes that our readers will join and also participate in progressing our common field of study.

So as you can see, we try to bring many diverse perspectives to the table, while making the pursuit of RCA a practical one, not a complex one. We certainly want to avoid falling into the "paralysis-by-analysis" trap when looking at something like RCA—that would be hypocritical, would it not?

We bring to light the perspectives of the pragmatic "rooticians" to the "purists" so that readers can make their own judgments as to what is best for their applications. We present debates on definitions of words commonly used in the RCA lexicon, but ultimately come to the conclusion that there are no generally accepted definitions in the field so we must fend for ourselves (which is part of the problem with communication).

There are many RCA methodologies on the market and we discuss them in generalities so as not to put the microscope on any individual or proprietary approach. In this manner we can discuss the pros and cons of each type of approach, and readers can decide the level of breadth and depth that they require in their analysis.

We discuss the scope of RCA—where does it begin and where does it end? How does a true RCA effort integrate with the organizational structure and remain a viable and valuable resource to the organization? Where there is RCA there is turf politics, so we discuss how this activity called RCA fits with existing initiatives like Total Quality Management (TQM), Reliability Engineering (RE), Reliability Centered Maintenance (RCM), and Six Sigma.

Our intent with this edition of this text is to expand the various perspectives brought to light on the topic of RCA and to present a current "state-of-the-RCA field" so that readers can make their own sound judgments as to how they wish to design and define RCA for their own organizations.

Will everybody who reads this text agree with its content? No. Can they benefit regardless? Yes. We hope to spark debate within the minds of our readers where the differences are contrasted between how we approach RCA and how they are currently conducting them at their facilities.

Perhaps we will sway some to agree with certain premises in this text, and others will improve upon their current approaches with the ideas presented. Either way, the journey of the learning is what is most important. Analysts will collect the necessary data, sift out the facts, and make their own determination as to what they believe is best for them.

Robert J. Latino
Kenneth C. Latino
Mark A. Latino

Reflections

In a divergence from the typical formatting of a book "Foreword," we wanted to use this introduction to reflect on the career of our father, Charles J. Latino, who made the successes of the Reliability tools described in this text possible over 40 years ago.

Charles J. Latino
Father of Manufacturing Reliability
1929–2007

Charles was known as the "Father of Manufacturing Reliability" as he founded and directed the first corporate Reliability Engineering Department in heavy manufacturing back in 1972 as a Research and Development (R&D) arm of Allied Chemical Corporation (more commonly known as Honeywell today). Charles was a visionary as he tried to introduce these Reliability principles at a time of economic prosperity in the United States. There were nothing but hurdles in attempting to ingrain a Reliability culture at a time when the organization could sell as much as it could make no matter what the quality of the product. Charles knew these prosperous times would not last forever. He knew there was no need to wait for equipment, processes, and human beings to break down before having to address their consequences.

He knew technologies existed that could allow his team to predict the signals of impending failure.

Charles persevered in influencing the corporation to establish a Corporate Reliability Engineering Department for all of Allied's holdings worldwide. This book is dedicated to Charles J. Latino (1929–2007) and his legacy. The following reflections are from members of Charles' original Reliability Engineering Team back in 1972, close colleagues, and family members as they reflect on the impact he and his Reliability ideas had on them as individuals and their subsequent careers. As you will learn, Reliability is not a program but a way of life.

JAN B. SMITH, P.E., ELITE MEMBER OF THE ORIGINAL ALLIED CHEMICAL CORPORATE RELIABILITY ENGINEERING GROUP

In the late 1960s, Charles was responsible for engineering and maintenance at Allied Chemical's Chesterfield nylon plant. The plant experienced a prolonged period of severe operational problems that the industry referred to as a "blitz." Continued operation at this level of performance was impossible. An option being considered by corporate to rectify the poor reliability and resulting process upsets and low yields was to increase equipment redundancy. Charles knew there was a fundamentally better and more cost-effective way to control reliability. He acted upon the opportunity that came with a difficult problem and convinced the plant manager and corporate management to try condition monitoring, failure prediction, and root cause failure analysis. As a result, Reliability Engineering that is practiced in process plants today was born. I was fortunate to be involved with this early work, made possible by Charles's vision and persuasion, and to be mentored by him for nearly 40 years.

After Charles took Allied's Reliability Center independent, I had many opportunities to work with him, his sons, and his staff. Charles led plant reliability studies for various clients, which gave me a chance to work side by side with him as a team member. Discussing identified reasons for nonoptimum plant performance, their solutions and strategies for their acceptance and implementation gave me insight into Charles that I never realized earlier. As most performance issues are management related, we often had observations and recommendations that plant management would rather not hear. Charles never balked at telling it like it was. While consultants often lean too far toward giving the client what the client wants to hear, Charles never had that inclination. He always pushed the client toward what the client needed for performance excellence. This allowed me to see his integrity and character firsthand. I have often heard character defined as "doing what is right when no one is looking," but Charles redefined character for me as "doing what is right, although others are looking and disapproving." I will forever be thankful to Charles for showing me what integrity and character truly look like. Because Charles was the man that he was, facilities around the world are more reliable and safer. But what is more important to me is that I am much better having had Charles as a friend, mentor, and example.

NEVILLE SACHS, P.E., ELITE MEMBER OF THE ORIGINAL ALLIED CHEMICAL CORPORATE RELIABILITY ENGINEERING GROUP

I'm not positive of when I met Charles, but I know he was responsible for my new job in 1973 and my continuing career 35+ years later. When I started that "new job" at Allied Chemical, the title was "Maintenance Engineer" and a few months later we became "Reliability Engineers."

In those days, the traditional engineering departments in chemical plants involved Plant Engineers, responsible for designing and installing the machinery and facilities, and Process Engineers, responsible for the chemical process and process equipment. In looking at the gaps between the true plant capacity and the typical production rates, Charles realized that there was a glaring discontinuity between the capabilities of the traditional structure and the needs of the plant. From this realization, he developed the idea of a multidiscipline "Reliability Engineering Department" that would work closely with the plant personnel to improve operating reliability. Allied Chemical formed Reliability Engineering Departments in several of their larger plants and we were the disciples trying to put "Charles's ideas" into practice.

The basic approach was to look at *everything and anything* that acted as a limit on the plant's capacity and then find ways to overcome it, and the term "Charles's ideas" can't begin to describe the scope of the projects. The initial investigations searched for physical failure causes, but they rapidly grew into in-depth analyses to uncover the latent roots, and the projects ranged from the development of routine predictive maintenance programs, to implementing specialized NDT practices, to installing a receiving inspection program for maintenance materials.

Charles often said that an investment into true root cause analysis returned more than one thousand fold. The results from these early projects were spectacular and helped the "Reliability Approach" to expand from Maintenance into Engineering and other areas of the corporate structure.

After Allied-Signal decided to change its corporate direction and divorce the corporation from the heavy chemicals business, Charles took the Reliability Center private and continued to preach the value of extending "failure analysis" into true "Root Cause Analysis" and the exposure of the latent roots. During this time, I had the pleasure of working with Charles (and Jan Smith) both in presenting many seminars on "The Reliability Approach" and in investigations into plant and corporate root causes.

I have always thought of Charles as humorous, honest, and a politically fearless gentleman. It was inevitable that, when these investigations found structural weaknesses in the way a plant or a corporation was operated, it was Charles's duty to tell them about their errors, and he politely told them "like it was," clearly explaining the logic behind the findings … and more than once was told not to come back.

Like all of us, he had weaknesses (he routinely made our seminar schedules an exciting challenge!), but he was a pleasure to work with, a great teacher, and a pioneer in transforming the way we think of maintenance and problem solving.

ED GOLL, OPERATIONS RESEARCH ENGINEER/CHANGE MANAGEMENT CONSULTANT, ELITE MEMBER OF THE ORIGINAL ALLIED CHEMICAL CORPORATE RELIABILITY ENGINEERING GROUP

Charlie was a man of courage and devotion. To me, he was a mentor, a friend, and sometimes a boss. Most memorably, however, he stood as a paragon example of what it meant to be committed to a vision and to hold true to that vision through thick and thin. His enthusiasm was contagious and his encouragement made anything seem possible and worth striving for. It was an honor and a blessing to have known and worked alongside Charlie.

C. ROBERT NELMS, AEROSPACE ENGINEER, ELITE MEMBER OF THE ORIGINAL ALLIED CHEMICAL CORPORATE RELIABILITY ENGINEERING GROUP

We all can count on one hand the number of people who have made profound influences on our lives. I recall, in 1994 when writing my first book, the struggle within me as I tried to decide whom to dedicate it to—my wife, my parents, or Charles J. Latino. I met him in 1974, and quickly knew him as Charlie—that's what we all called him. In the early years, I was in awe of him—his insights, his mannerisms. He was monumental. Until I met Charlie, I was "timid and safe." But through the years, as I came to know him better, his impact on me can be summed up in the words of the book he had wanted to write: "Bold and Outrageous." He found that path and cleared the way. I chose to follow. Later on, when he first established the present-day Reliability Center, I had the honor of traveling alone with him all over the United States and Canada, helping to carry his bags, get medicines for him, share so many private dinners, go on walks with him while he recuperated from illnesses. He became part of me. Looking back, I realize that I am who I am, to a huge extent, because of Charles Latino.

GEORGE PATE, RELIABILITY ENGINEER, ELITE MEMBER OF THE ORIGINAL ALLIED CHEMICAL CORPORATE RELIABILITY ENGINEERING GROUP

I first met Charles (Charlie) Latino in December of 1963. He was the head of the Plant Engineering Department at an Allied Chemical (now Honeywell) Plant in Chesterfield, VA. I was a college senior interviewing for an engineering position upon my graduation in June of 1964. Charlie offered me a job and thus became my boss, mentor, and friend.

During the subsequent years, I witnessed Charlie develop and implement programs to improve the mechanical reliability of this plant. Recognizing the benefits of these programs, Charlie led efforts to install these programs in other plants in the corporation. Ultimately Charlie decided to form Reliability Center, Inc. to share his concepts with other corporations throughout the world. He became a much sought-after speaker at trade conferences and seminars globally.

As the years went by, my friendship with Charlie grew. Along with the friendship, our trust in one another grew also. Charlie would sometimes get my input on very important decisions that he had to make for the company. He and his wife, Marie, and my wife, Linda, and I would sometimes vacation together. We lost money at the casino and horse tracks, laughed a lot, ate at fine restaurants, drank good wine, and had a great time together.

AL THABIT, RELIABILITY ENGINEER, ELITE MEMBER OF THE ORIGINAL ALLIED CHEMICAL CORPORATE RELIABILITY ENGINEERING GROUP

In 1973, straight out of college, I became a member of Allied Chemical's reliability group under the leadership of Charles (Charlie) Latino. He always introduced himself as Charles but we all called him Charlie.

Those were heady days. We worked on the cutting edge of concepts and technologies that changed how we solved problems and how maintenance was performed. I remember thinking at the time, "This is how maintenance is supposed to be done," but when I would try explaining to colleagues what we were doing I would get blank stares in response. It was that revolutionary! Charlie was a champion of this change. He took on nonbelievers all the way up the corporate ladder.

In the mid 1990s I convinced Charlie to come to a gold mine in Nevada where I worked to give a seminar on Reliability Centered Maintenance. We spent 4 days working long hours, but his enthusiasm for the subject remained strong and contagious. It was during this time that Charlie introduced me to his term "bold and outrageous" in regard to getting these concepts accepted. It is a term I understood immediately based on how Charlie worked to convince people of the value of reliability concepts throughout his career. It is how those of us who worked for him have tried to live our lives.

Charlie was my mentor and a true friend. He taught me a great deal that has served me well throughout my career.

GARY LEE, RELIABILITY ENGINEER, ELITE MEMBER OF THE ORIGINAL ALLIED CHEMICAL CORPORATE RELIABILITY ENGINEERING GROUP

In 1974 I was just out of business school with a newly minted MBA when I joined the Reliability Center. Most of my "B" school contemporaries were joining banks and brokerage firms and questioned my decision to join an engineering organization. But I had good reason. I had family connections to Allied Chemical (now Honeywell) and they told me Charlie Latino was making big waves within Allied. And as any MBAs will tell you, working for a wave maker is usually an exciting and beneficial place to be. I was right to join the Reliability Center because I quickly confirmed that although Charlie was an engineer, he was foremost a gifted leader and businessman who was determined to dramatically improve the efficiency of Allied's far-flung manufacturing processes. Charlie proved over and over again that Reliability Engineering was an effective bottom-line-oriented business tool.

Most everything I learned about leadership and integrity I learned from Charlie Latino. In the years since working for Charlie I have often formulated successful solutions by asking, "What would Charlie do?"

PAT WHELAN, RELIABILITY ENGINEER, ELITE MEMBER OF THE ORIGINAL ALLIED CHEMICAL CORPORATE RELIABILITY ENGINEERING GROUP

Shortly after the improper handling and dumping of Kepone, a carcinogenic insecticide, caused a nationwide pollution controversy and the closure of the James River to fishing between Richmond and the Chesapeake Bay, I had the opportunity to meet Charles J. Latino. Charlie had started a reliability engineering group out of Allied Signal Company's Hopewell, VA, manufacturing facility, and his department was engaged to assure that the processing facility for a new specialty chemical (i.e., hazardous material) operation was constructed and operated as engineered. I was so impressed with the Reliability Center's work and Charlie's leadership that I asked to join his team!

The next few years of working with Charlie and our team brought a continuous stream of new technical challenges where his insights, character, and competence not only made following him easy, but provided an invaluable leadership model that benefited all of us. On a daily basis, or whenever there was a crisis to be addressed, Charlie was smart enough to let us make the decisions we had to make, wise enough to know when to intervene, and shrewd enough to provide "air cover" so we could grow professionally.

Few individuals in our lives have such an impact … Charlie's lessons are with me always!

DON PICKUP, SUPERVISOR, RELIABILITY ENGINEERING, DELAWARE WORKS, CLAYMONT, DELAWARE

When a testing group became a "Reliability Engineering" department, what was different from the past? Mechanical failures still happened. Much of the difference was Charlie. It was no longer just repair or replace, it was time to determine the "root cause" of the failure, fix the root cause, and stop the failure. Eddy current testing, high-speed photography, and nuclear devices were added to vibration and thickness testing equipment. The failed bearing was studied. Maybe it's better to upgrade it, change the design, or change the material. Things ran better and longer. That's a good thing.

Charlie also looked at people, all people—mechanics and engineers, operators and chemists. Often now, looking at people is to place blame. Charlie looked at people to learn, to do things better, to do jobs better. Sometimes a change could reduce "human error." That's a good thing.

I don't know if Charlie ever used the term "think outside the box," but he believed it. Doing things as before may be all right, but if things can be done better, if things can be done easier, or if things can last longer, Charlie would insist that changes are needed. I miss Charlie Latino.

KENNETH C. LATINO, CMRP, RELIABILITY CHAMPION, MEADWESTVACO (SON OF CHARLES J. LATINO)

My father was a great inspiration to me. I had the honor of being able to work with him for nearly 15 years at the Reliability Center. As a boss, he was sometimes tough and demanding but always fair and with purpose. At the time, I did not always care for those traits but as the years have passed, his passion and work ethic have stuck with me. I often look back and think about how my dad would have handled this or that situation. For that gift, I am forever grateful.

My father was truly the pioneer in the field of Reliability Engineering. His vision and passion for excellence were shared by literally thousands of people throughout the world through his many writings and teachings. Much of what you see in any industrial plant today evolved from his vision of what a reliable industrial plant could be.

My father left a tremendous legacy to the world that my family carries on today. They continue to communicate Charles Latino's passion for Reliability through their work at the company he created more than a quarter-century ago. Although I miss my father every single day, I know that his great work lives on in every corner of the globe. I hope that you will enjoy reading and learning from this book. You can be sure that his influence permeates every facet of this text. Enjoy!

About the Authors

Robert J. Latino is CEO of Reliability Center, Inc. (RCI). RCI is a reliability consulting firm specializing in improving equipment, process, and human reliability. He received his bachelor's degree in business administration and management from Virginia Commonwealth University.

Robert has been facilitating RCA and FMEA analyses with his clientele around the world for over 20 years and has taught over 10,000 students in the PROACT® methodology. He is co-author of numerous seminars and workshops on FMEA and RCA as well as co-designer of the award winning PROACT Suite Software Package.

Robert is a contributing author of *Error Reduction in Healthcare: A Systems Approach to Improving Patient Safety* and *The Handbook of Patient Safety Compliance: A Practical Guide for Health Care Organizations.*

Robert has also published a paper entitled, "Optimizing FMEA and RCA Efforts in Healthcare" in the *ASHRM Journal* and presented a paper entitled, "Root Cause Analysis Versus Shallow Cause Analysis: What's the Difference?" at the ASHRM 2005 National Conference in San Antonio, Texas. He has been published in numerous trade magazines on the topic of reliability, FMEA, and RCA and is also a frequent speaker on the topic at domestic and international trade conferences.

Robert has also applied the PROACT methodology to the field of Terrorism and Counter Terrorism via a published paper entitled, "The Application of PROACT RCA to Terrorism/Counter Terrorism Related Events."

Kenneth C. Latino is Reliability Champion of MeadWestvaco in Covington, Virginia. He has a bachelor of science degree in computerized information systems from Virginia Commonwealth University. He began his career developing and maintaining maintenance software applications in the continuous process industries. After working with clients to help them become more proactive in their maintenance activities, he began consulting and teaching industrial plants how to implement reliability methodologies and techniques to help improve the overall performance of plant assets.

Over the past few years, a majority of Kenneth's focus has centered around developing reliability approaches with a heavy emphasis on Root Cause Analysis (RCA). He has trained thousands of engineers and technical representatives on how to implement a successful RCA strategy at their respective facilities. He has co-authored two RCA training seminars: one for engineers and another for hourly personnel.

Kenneth is also co-software designer of the RCA program entitled the PROACT Suite. PROACT was a National Gold Medal Award winner in *Plant Engineering's* 1998 and 2000 Product of the Year competition for it first two versions on the market. He is currently president of the Practical Reliability Group, a Reliability consulting firm dedicated to delivering approaches and solutions that can be practically applied in any asset-intensive industry.

Mark A. Latino is president of Reliability Center, Inc. (RCI). Mark came to RCI after 19 years in corporate America. During those years, a wealth of reliability, maintenance, and manufacturing experience was acquired. He worked for Weyerhaeuser Corporation in a production role during the early stages of his career. He had an active part in Allied Chemical Corporation's (now Honeywell) Reliability Strive for Excellence initiative that was started in the 1970s to define, understand, document, and live the Reliability culture until he left in 1986. Mark spent 10 years with Philip Morris primarily in a production capacity that later ended in a Reliability Engineering role. Mark is a graduate of Old Dominion University and has a bachelor's degree in business that focused on production and operations management.

1 Introduction to the PROACT® Root Cause Analysis (RCA) Work Process

Effective RCA can arguably be one of the most valuable tools to any organization. This is especially true for large, asset-intensive companies. There are many issues that arise, and if there is not a plan in place to deal with these issues, the facility can become very reactive.

The challenge with effective RCA is when do we apply the resources to identify the root causes of a problem? There are simply too many issues that arise to effectively solve every one. Therefore, a more intelligent approach must be taken to select the right issues to resolve.

Let's look at a simple example. Let's assume that we have two centrifugal pumps. One of the pumps is a charge pump that is critical to the operation of the unit it serves. The other is a water pump that is spared and is not deemed in a critical service. Which problem do we analyze if we are experiencing problems with both of these pumps and there are limited resources to address the root causes? The critical charge pump, of course.

We often see organizations struggle with which failures to analyze using RCA. Very often, analysis work is limited to regulatory issues like safety and environmental events. Many times, equipment or process-related issues are simply corrected and the process is started back up without knowing the cause. Without identifying and addressing the various root causes, the problem is likely to recur. It seems that without some sort of outside pressure to perform an analysis it simply does not happen. Therefore, a strategy should be employed to direct personnel on what and when to do RCA.

As previously stated, there are many issues that occur on a daily basis at a large, asset-intensive facility. When these issues occur they are deemed very important and must be addressed. We need some way to separate the emotion of the "failure-of-the-day" to what is truly important to the success of the facility. Therefore, we need to determine what the perspectives, objectives, and measures are for the organization. For example, perhaps your plant has a mandate to improve profitability without the expenditure of additional capital. How would you go about doing that? You need a strategy to determine what you are going to do and how you are going to measure it.

We work in a lot of facilities that are measuring many things related to their operation. Many organizations develop the metrics that they feel are important to measure as they progress into a maintenance and reliability initiative. We often hear about Mean-Time-Between-Failures (MTBF), Mean-Time-to-Restore (MTTR), and many other measurements. Measuring performance for the sake of measuring is not especially useful unless the measurements are directly related to the performance of the organization and action is taken to make the needed improvements when the measures are going in a negative direction.

Therefore, we must first think about what goals or objectives we are trying to accomplish before we can determine what measures we need to monitor. An effective methodology for determining your company's objectives is to create a *strategy map*. A strategy map takes all of the objectives of the company and puts them into various perspectives. The perspectives can vary from company to company, but for the area of asset management there are four main perspectives:

1. Corporate
2. Assets
3. Work Practices
4. Knowledge and Experience

Within each of the four perspectives, a number of individual objectives are defined. For instance, within the Corporate Perspective we look at objectives that directly relate to goals defined within the company. These are typically related to the fiscal performance of the business but can also be related to critical operational issues like environmental and safety performance. Other objectives related to the Corporate Perspective might be customer satisfaction issues like on-time deliveries, quality of the product, and many others. However, in the area of asset management we typically focus on those areas that relate to financial, safety, and environmental performance as it relates to the utilization of assets.

Following is a list of typical perspectives and objectives related to asset management:

1. Corporate Perspective
   a. Increase Return on Investment (ROI)
   b. Improve Safety and Environmental Conditions
   c. Reduction of Controllable Lost Profit
   d. Reduction of Maintenance Expenses
   e. Increase Revenue from Assets
   f. Reduce Production Unit Costs
   g. Increase Asset Utilization
   h. Minimize Safety and Environmental Incidents
2. Asset Perspective
   a. Minimize Unscheduled Equipment Downtime
   b. Improve System Availability
   c. Reduce Scheduled Maintenance Downtime
   d. Reduce Unscheduled Repairs

   e. Reduce Non-Equipment-Related Downtime
   f. Increase Equipment Reliability
   g. Reduce Equipment Failure Time
3. Work Practices Perspective
   a. Reduce Repair Time
   b. Reduce Maintenance Material Inefficiencies
   c. Improve Labor Efficiency
   d. Improve Material Purchasing
   e. Perform Predictive Maintenance
   f. Optimize Time-Based Maintenance
   g. Optimize Work Processes
   h. Perform Reliability Studies
   i. Perform Criticality and Risk Assessments
   j. Improve Maintenance Planning and Scheduling
4. Knowledge and Experience Perspective
   a. Improve Historical Equipment Data Collection
   b. Improve Operations Communications
   c. Train Maintenance and Operations Personnel

Once the perspectives and objectives are fully defined we need to determine the relationship of lower-level objectives to upper-level objectives. Figure 1.1 shows an example of a sample strategy map with the objective relationships defined for the Corporate Perspective.

Strategy maps are effective visual vehicles for demonstrating how every person in the organization can affect the performance of the overall business. For instance,

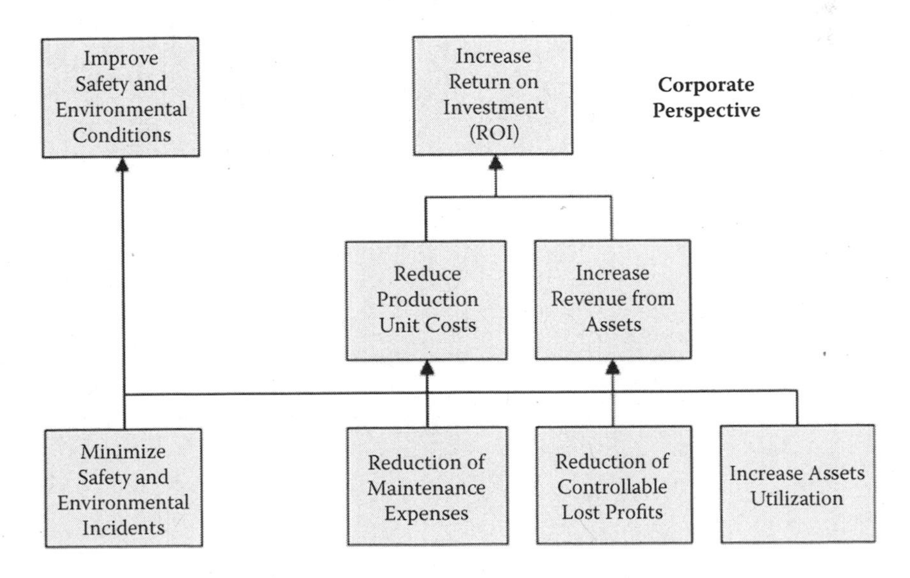

FIGURE 1.1 Sample Corporate Perspective strategy map.

when a technician is performing vibration analysis in the field, he or she can see how the application of that skill will improve equipment reliability. This will ultimately contribute to the corporate goals of achieving higher returns on the capital that has been employed.

Let's return to the concept of metrics and Key Performance Indicators (KPIs). Tom Peters once said, "You can't improve what you cannot measure." If you think about it for a minute it makes a lot of sense. We are exposed to KPIs starting when we are very young. From the moment we are born we are weighed and measured and then we are compared to standards to see in which percentile we are. As we grow and get into school we are exposed to another set of KPIs—the infamous report card. The report card allows us to compare our performance against our peers or to some standard. An example that many people can certainly relate to is the use of a scale to measure the progress of a diet. We probably would not be very successful if we did not know where we started and what progress we were making week by week.

We all need a "scoreboard" to help us determine where we started and where we are at any given time. This certainly applies to measuring the performance of a maintenance and reliability organization. We need to know how many events occur in a given month, on a specific class of equipment, etc. Not until we know what KPIs will effectively measure our maintenance and reliability objectives can we begin to establish which opportunities will afford the greatest returns.

With all of that said, we would like to provide a word of caution. Be very careful to diversify your KPI selections. While a report card in school is a good measure of a student's performance, it still does not provide a complete picture of the individual student. It is only one data point! Some students perform better on written tests while other students excel in other ways. We need to be careful to make sure that we employ a set of KPIs that most accurately represents our performance. That means having many different metrics that look at different areas of performance so we can get a complete picture.

So let's take a look at a few common Reliability KPIs that can be employed to give us an understanding of our overall asset performance.

MTBF (MEAN-TIME-BETWEEN-FAILURES)

This is a common metric that has been used for many years to establish the average time between failures. Although it can be calculated in different ways, it primarily looks at the total runtime of an asset(s) divided by the total number of failures for that asset(s).

$$\text{Total Runtime/Number of Events} = \text{MTBF} \tag{1.1}$$

EQUATION 1.1 Sample MTBF calculation.

This is a good metric because it is easy for people to understand and relate to and is common throughout industry.

NUMBER OF EVENTS

This metric will simply measure the volume of events that occur in a variety of dimensions. Those dimensions are typically process units, equipment classes (e.g., pumps), equipment types (e.g., centrifugal pumps), manufacturer, and a host of others. This metric is closely related to MTBF as it is the denominator for the calculation. It can also be an accurate reflection of a facility's maintenance and reliability performance.

MAINTENANCE COST

This metric simply measures the number of maintenance dollars that are expended on rectifying the consequence of an event. This is typically the sum of labor and material costs (including contractor costs). This metric is also employed across many different dimensions like equipment, areas, manufacturers, etc. It is a better business metric as it shows some of the financial consequences of the event. It also has some drawbacks, as it does not reflect the complete financial consequence of the event, and it does not cover the lost opportunity (e.g., downtime) associated with the event. As we all know, the cost of downtime is much greater than the cost of maintenance in a dramatic downtime event.

AVAILABILITY

This metric is useful to determine how available a given asset or set of assets has been historically. In a 24/7 operation the calculation is simply the entire year's potential operating time minus downtime divided by total potential operating time. Equation 1.2 is a Sample Availability calculation.

$$\frac{8760 \text{ (total hrs. in a year)} - 32 \text{ (4 failures of 8 hours each)}}{8760 \text{ (total hrs. in a year)}} \tag{1.2}$$

$$\textbf{Availability} = 99.63\%$$

This calculation can be modified in many ways to fit a specific business need. Although this metric is a good reflection of how available the assets were in a given time period, it provides absolutely no data on the Reliability or business impact of the assets.

RELIABILITY

This metric can be a better reflection of how reliable a given asset(s) is based on its past performance. In the availability example above, we had an asset that failed four times in a year, resulting in 32 hours of downtime. The availability calculation determined that the asset was available 99.63% of the time. This might give the impression of a highly reliable asset. But if we use the Reliability calculation that follows we would get a much different picture. Equation 1.3 is a Sample Reliability calculation.

$$\begin{aligned}
&\text{Reliability} = e^{-\lambda t}\\
&\text{Natural logarithmic base: } e = 2.718\\
&\text{Failure rate: } \lambda = 1/\text{MTBF} = 1/91\\
&\text{Mission time: } t = 365 \text{ (days)}\\
&\text{------------------------------}\\
&\text{Reliability} = e^{-\lambda t}\\
&\qquad = 2.718^{-\lambda t}\\
&\qquad = 2.718 - 1/91\ (365)\\
&\qquad = 2.718^{-4.0109}\\
&\qquad = 1.81\%
\end{aligned} \tag{1.3}$$

The fact is, an asset that fails four times per year is extremely unreliable and the likelihood of that asset reaching a mission time of 1 year is highly unlikely, even though its availability is very good.

These are only a few common KPIs. As you can imagine, there is an array of metrics that can be used to help measure the effectiveness of a maintenance and reliability organization. We will discuss these in more detail in just a moment.

So we now understand that MTBF, MTBR, Availability, and many other measures are commonly used to determine the effectiveness of equipment reliability. But unless these metrics are measuring the performance of a given company objective, they might not provide the benefit that the company is trying to achieve. Therefore, we need to first look at each objective and then develop pertinent measurements to see if that objective is indeed being met.

For example, if our objective were to reduce production unit costs, we would measure the cost per unit of product produced. This will help us to understand if we are getting better, worse, or staying constant with respect to our production costs. However, this alone is not enough. We need to be more specific when we are defining our measurements. The Key Performance Indicator or KPI, as it is often called, needs to delineate the difference between good and poor performance. For example, let's assume that our average cost per unit of product is $10 this month. Is that cost high, medium, or low? In order to have an indicator, you must define the measurement thresholds. In our example, we said that the average cost per unit this month was $10. Perhaps our target value for production unit cost is $8. Therefore, our performance is not very good.

A KPI has several thresholds that should be defined prior to the monitoring of the measure's value. These are

1. Target Value—This value specifies the performance required to meet the objective.
2. Stretch Value—This value represents performance above and beyond what is expected to meet our objectives.

3. Critical Value—This value represents performance that is deemed unacceptable for meeting our objectives.
4. Best Value—This is the best possible value for this objective.
5. Worst Value—This is the worst possible value for this objective.

When these thresholds are set properly for each measurement, we can objectively assess our performance. Otherwise, we are simply collecting information with no real sense of whether the value is meeting our specified goals.

Let's get back to the strategy map discussion. The process is to review each objective that we deem as important to our strategy and list one or more KPIs that will be accurate measurements for that objective. Once we define the measurement and calculation, we need to determine the target, stretch, critical, best, and worst values for that measure. Upon completion of this process we will have a completed strategy map. Table 1.1 lists some example KPIs that relate to our objectives and perspectives.

BALANCED SCORECARD

Let's explore the process of monitoring KPIs on a routine basis. We will employ a *balanced scorecard* methodology to help us do this. A balanced scorecard takes the perspectives, objectives, and measures introduced in the strategy map and puts them into an easily understandable format. A sample of a balanced scorecard and KPI measurement is displayed in Figures 1.2 and 1.3.

Having all of your critical performance information displayed in one place makes it easy for everybody involved in the enterprise to see their performance and to determine where to focus their attention.

This process will ensure that we are working on the critical issues that most affect the performance of the business. Once we begin to monitor the balanced scorecard on a routine basis we will begin to see the areas where we need to make improvements. For example, let's say we are monitoring unscheduled downtime as a measurement of the equipment downtime objective. We observe that our performance for that KPI is well below the target level. We then must investigate and collect information to see which events are contributing to the poor performance for that objective.

THE RCA WORK PROCESS

A successful RCA initiative must have a strategic and tactical plan in place (Figure 1.4). We discussed the concept of a strategy map to ensure that we are measuring the key metrics that will enable us to achieve our company's objectives. Let's talk more about the tactical plan for implementing the RCA initiative.

First of all, we must have a means of collecting data related to the events that affect the performance of our stated objectives. This can be maintenance data, process data, and other data related to the performance of our facility. We will talk much more about event data collection in Chapters 4, 5, and 8.

Once we have a process for collecting data on these events, we must decide on criteria that will initiate the execution of an RCA analysis. For example, your strategy might dictate that if any failures occur on critical equipment, an RCA must be

TABLE 1.1
Sample Completed Strategy Map

Perspective Description	Objective Description	KPI Description
Corporate Perspective	Improve Safety and Environmental Conditions	Number of overall safety and environmental incidents
	Increase Asset Utilization	Overall equipment effectiveness
	Increase Asset Utilization	Utilization rate by unit %
	Increase Asset Utilization	Plant utilization
	Increase Return on Investments (ROI)	Return on Capital Employed (ROCE)
	Increase Revenue from Assets	Production throughput
	Minimize Safety and Environmental Incidents	Safety and environmental incidents
	Minimize Safety and Environmental Incidents	Accident by type, time of day, craft, personnel age, training hours attended, supervisor, unit, area
	Reduce Production Unit Costs	Cost Per Unit
	Reduction of Controllable Loss Profit	Lost profit opportunity cost
	Reduction of Maintenance Expenses	Annual maintenance cost/Asset replacement cost
	Reduction of Maintenance Expenses	Maintenance cost
	Reduction of Maintenance Expenses	WO Cost, 2-Mo-Avg
	Reduction of Maintenance Expenses	Cost of PM by equipment type
	Reduction of Maintenance Expenses	Maintenance cost per barrel of product produced
	Reduction of Maintenance Expenses	Cost of PdM by equipment type
	Reduction of Maintenance Expenses	Unplanned cost as a % total maintenance cost
Asset Perspective	Improve System Availability	Unit availability
	Improve System Availability	Uptime
	Improve System Availability	Onstream factor
	Improve Equipment Reliability	Average cost per repair
	Improve Equipment Reliability	MTBR
	Improve Equipment Reliability	MTBF
	Minimize Unscheduled Equipment Downtime	Number of Lost profit opportunity events
	Reduce Equipment Failure Time	Equipment failure downtime
	Reduce Non-Equipment Downtime	Downtime due to quality, feedstock, production scheduling

TABLE 1.1 (continued)
Sample Completed Strategy Map

Perspective Description	Objective Description	KPI Description
	Reduce Scheduled Maintenance Downtime	Turnaround downtime
	Reduce Unscheduled Repairs	Number of failures
	Reduce Unscheduled Repairs	% of emergency repairs
Work Practice Perspective	Improve Labor Efficiency	Labor cost of repairs
	Improve Maintenance Planning and Scheduling	% of emergency (break in) work orders
	Optimize Time-Based Maintenance	% of critical equipment with PM optimized
	Optimize Work Processes	% of rework
	Optimize Work Processes	Hours of overtime
	Optimize Work Processes	% of overdue work orders
	Perform Criticality and Risk Assessment	Number of failures on critical and high risk equipment
	Perform Predictive Maintenance	% of PdM generated work
	Perform Reliability Studies	Number of new work orders generated from Reliability Analysis
	Reduce Maintenance material Inefficiencies	Average parts wait time
	Reduce Repair Time	MTTR
Knowledge and Experience Perspective	Improve Operations Communications	Number of defects observed from operators
	Improve Historical Equipment Data Collection	% of populated required fields in work order history
	Train Maintenance and Operations Personnel	Hours of training per employee
	Train Maintenance and Operations Personnel	Dollars spent on training per employee

performed. This is very common for events that relate to safety and environmental performance. We do not want to leave this process too ambiguous because people will not know when and under what circumstances to conduct an analysis.

It may be that you want to employ different levels of analysis for different performance criteria. Perhaps you have many events that occur on noncritical equipment, but the frequency of the events is causing a large amount of maintenance expenditure. This might not justify a full-blown team to perform the analysis but would still justify some level of analysis to determine the reasons for the chronic maintenance

Sitemap: Metrics->Scorecard->Maintenance and Reliability Scorecard

View Maintenance and Reliability Scorecard

	Actual	Previous	Target	Trend	Frequency	Measurement Date
Corporate						
Increase Return on Investment (ROI)						
Return on Capital Employed (ROCE)	21.00	26.00	25.00	↓	Quarterly	6/1/2005
Improve Safety and Environmental Conditions						
Number of overall safety and environmental incidents	1.00	2.00	2.00	↓	Quarterly	6/1/2005
Reduction of Controllable Lost Profits						
Lost Profit Opportunity	150000.00	175000.00	200000.00	↓	Monthly	8/1/2005
Asset						
Minimize Unscheduled Equipment Downtime						
Number of Lost Profit Opportunity Events	5.00	1.00	3.00	↑	Monthly	8/1/2005
Improve System Availability						
Unit Availability	94.10	94.80	97.00	↓	Monthly	8/1/2005
Reduce Scheduled Maintenance Downtime						
Turnaround Downtime Days	13.00	5.00	10.00	↑	Quarterly	6/1/2005
Work Practices						
Reduce Repair Time						
MTTR	11.00	9.00	4.00	↑	Monthly	8/1/2005
Reduce Maintenance Material Inefficiencies						
Average Parts Wait Time	1.20	1.40	1.00	↓	Monthly	8/1/2005
Knowledge and Experience						
Improve Historical Equipment Data Collection						
% of populated required fields in work order history	71.00	76.00	80.00	↓	Monthly	8/1/2005

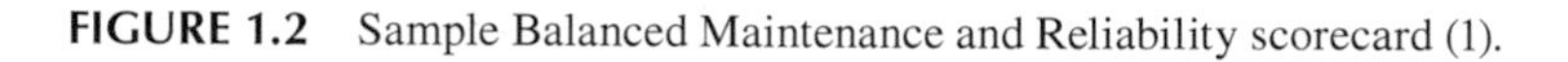

FIGURE 1.2 Sample Balanced Maintenance and Reliability scorecard (1).

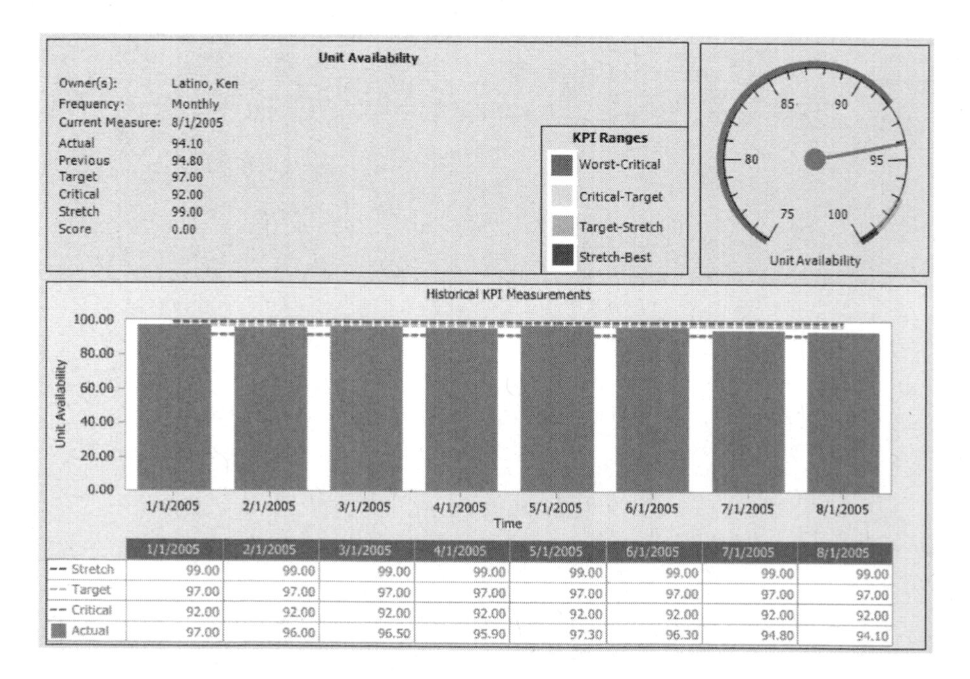

Unit Availability

Owner(s):	Latino, Ken
Frequency:	Monthly
Current Measure:	8/1/2005
Actual	94.10
Previous	94.80
Target	97.00
Critical	92.00
Stretch	99.00
Score	0.00

	1/1/2005	2/1/2005	3/1/2005	4/1/2005	5/1/2005	6/1/2005	7/1/2005	8/1/2005
Stretch	99.00	99.00	99.00	99.00	99.00	99.00	99.00	99.00
Target	97.00	97.00	97.00	97.00	97.00	97.00	97.00	97.00
Critical	92.00	92.00	92.00	92.00	92.00	92.00	92.00	92.00
Actual	97.00	96.00	96.50	95.90	97.30	96.30	94.80	94.10

FIGURE 1.3 Sample Balanced Maintenance and Reliability scorecard (2).

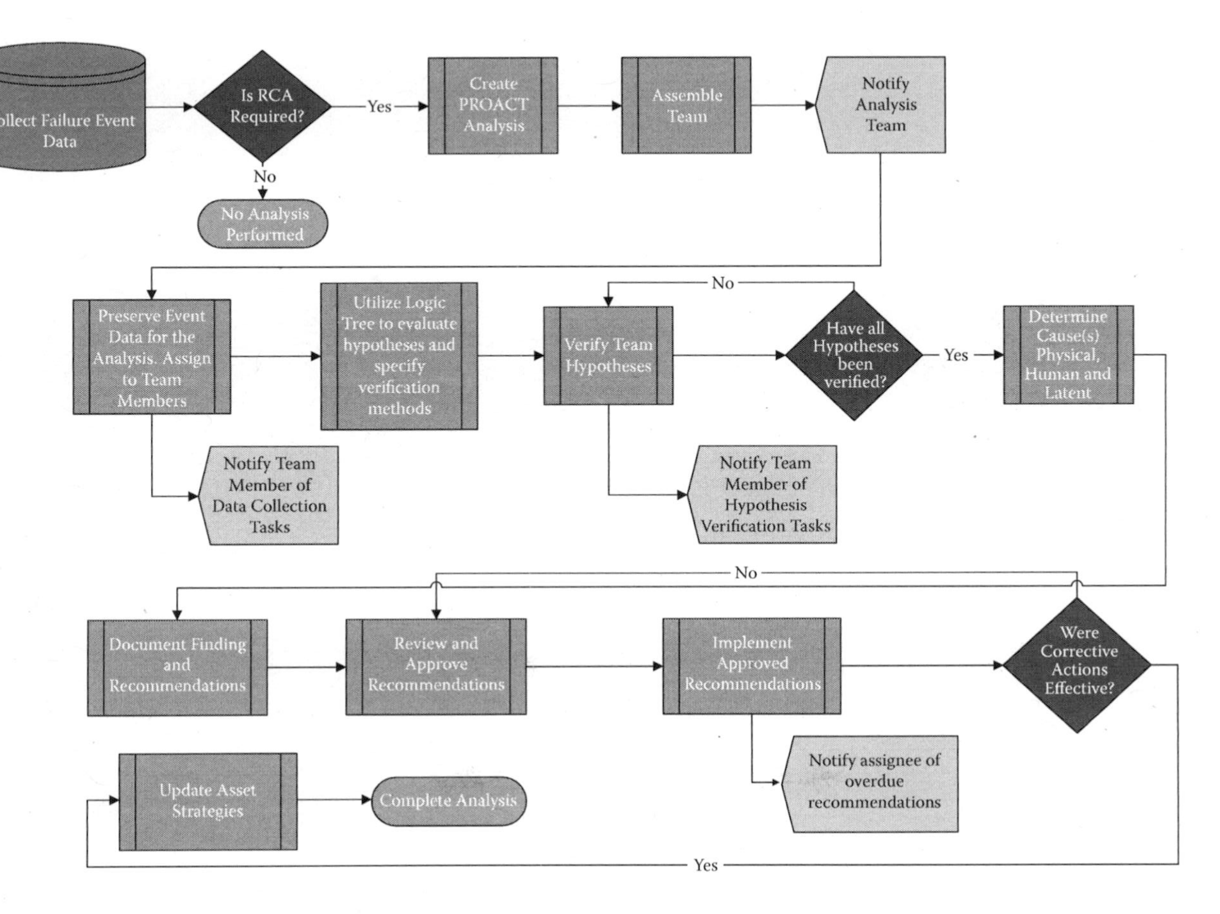

FIGURE 1.4 Sample RCA Work Process.

events. These types of analyses might be much less formal than a full-blown RCA but are still valuable.

Since every company is different and thus has different goals and objectives, it would not be prudent for us to define a generic criterion. However, we can delineate some examples that might be considered. In any plant, there is a need to optimize maintenance expenditures. Therefore, we may want to consider a criterion that is based on the amount of maintenance expended for a given piece of equipment for a fixed time period (e.g., the last 12 months). If a piece of equipment exceeds the threshold in that time period, an RCA will automatically be initiated.

Another common criterion can be production losses. This is especially true if your plant capacity is limited and you can market and sell everything that is produced in your facility. If there is a production loss that exceeds a specific financial value, an RCA should be initiated.

These are simple examples, but it is important to make sure that there is an agreed upon criteria for when RCA analyses will be initiated and who will perform the analyses. At many facilities, there is a Reliability Engineer responsible for a given area of the facility, and this person is responsible for performing RCAs on equipment/events in his or her area. It is then the Reliability Engineer's responsibility to determine what additional team members will be necessary to perform the analysis. We will discuss team formation in greater detail in Chapter 9.

The key to a successful analysis is to make sure that you have the data and subsequent information to determine what the underlying causes of the issue being studied are. The team will review the problem and determine what data will be needed to determine the root causes. The PROACT methodology offers a simple but effective acronym called the 5P's to help in this effort. The 5P's represent the five categories of data required to analyze any problem. We will discuss the data collection effort, and more specifically the 5P's, in Chapter 8.

Have you ever sat in a brainstorming meeting to solve a particular problem in a company? This is a very common approach to problem solving. We are not against the concept of brainstorming. In reality, we think it is a required activity in the RCA analytical process. The problem with most brainstorming sessions is that the group presents a variety of ideas but sometimes lacks the data to verify that the solution will work. For this reason, the PROACT methodology will utilize a Logic Tree approach to solve problems. This is a visual brainstorming tool. It is a hierarchical approach where the problem is defined in the beginning of the process, and subsequently hypotheses and verifications are formulated and proved out. The end goal of the process is to identify the true root causes of the problem. These causes can be physical, human, or latent in nature. We will discuss this in Chapter 10.

Identification of root causes, albeit important, will not solve the problem. The only way for the problem to be resolved is to implement corrective actions. This is typically done by creating a list of recommendations directed at eliminating or reducing the impact of the identified root causes. These must be thoroughly reviewed by all parties to ensure that they are the right solutions. Although causes are facts and cannot be disputed, recommendations should be thoroughly scrutinized and modified to ensure that they are the best course of action. We will discuss the process of communicating team findings and recommendations in Chapter 11.

As time passes we sometimes forget to follow up to make sure that our corrective actions were implemented and are providing the specified return we had intended. If the losses related to the problem are still affecting plant performance and negatively affecting our corporate strategy, then we should reevaluate our corrective actions to determine why they are not providing the intended benefit. The strategy map discussed previously will help, but we recommend having reevaluation criteria set for each recommendation. For example, we might measure the number of failures on that piece of equipment, and if another failure occurs in the next 12 months, we should reevaluate to see if the failure was related to the ineffectiveness of our corrective actions. We will discuss tracking results in Chapter 12.

Let's revisit our discussion on data collection methods. We have various methods to collect historical event information. We would like to break it into two categories—manual and automated data collection processes. In Chapter 5 we will discuss a process called Opportunity Analysis (OA) where we collect the data through the use of an interview process involving various personnel within the affected area. In the subsequent chapter we will discuss a more automated approach to data collection that will utilize existing information systems that may already be employed at the company.

There are pros and cons to both approaches. It generally comes down to data collection processes and how effectively they are employed. Many companies utilize a Computerized Maintenance Management System, or CMMS, to manage maintenance work and to document work history. These systems are often not utilized to their full potential, and many times work history on assets is not fully documented. If this is the case, then a manual interview process can be utilized to perform the opportunity analysis.

Now that we have explored the concept of the RCA Work Process, we will narrow the scope and look into the field of RCA itself and what it means in the industry, from both user and provider perspectives.

2 Introduction to the Field of Root Cause Analysis

WHAT IS ROOT CAUSE ANALYSIS (RCA)?

What a seemingly easy question to answer, yet no standard, generally accepted definition of RCA exists in the industry today, of which we are aware. Technical societies, regulatory bodies, and corporations have their own definitions, but it is rare to find two definitions that match. For the sake of having an anchor or benchmark definition, we will use the definitions provided by the Department of Energy (DOE) Guideline entitled "Root Cause Analysis Guidance Document (DOE-NE-STD-1004-92)."*

In the DOE document referenced above, the following is cited:

> The basic reason for investigating and reporting the causes of occurrences is to enable the identification of corrective actions adequate to prevent recurrence and thereby protect the health and safety of the public, the workers and the environment.

The document goes to say that every root cause investigation and reporting process should include the following five phases:

I. Data Collection
II. Assessment
III. Corrective Actions
IV. Inform
V. Follow Up

When we look at any investigative occupation, these five steps are critical to the success of the investigation. As we progress through this text, we will align the steps of the PROACT® methodology with each of these steps in the DOE RCA process.

For the purposes of this text, while aligning with the DOE guideline, we will use our own definition of Root Cause Analysis, which is

> The establishing of logically complete, evidence-based, tightly coupled chains of factors from the least acceptable consequences to the deepest significant underlying causes.

This is a variation of a definition that was proposed on the root cause analysis discussion forum at www.rootcauselive.com.†

---

* http://www.hss.energy.gov/nuclearsafety/ns/techstds/standard/nst1004/nst1004.pdf

† This discussion forum is associated with www.rootcauselive.com and moderated by Mr. C. Robert Nelms.

While a seemingly complex definition, let's break down the sentence into its logical components and briefly explain each:

1. Logically Complete—This means that all of the options (hypotheses) are considered and either proven or disproven using hard evidence.
2. Evidence Based—This means hard evidence is used to support hypotheses as opposed to using hearsay and treating it as fact.
3. Tightly Coupled Chains of Factors—This means that we are using cause-and-effect RCA approaches as opposed to categorical RCA approaches. We will discuss these differences in approach when comparing RCA tools.
4. Least Acceptable Consequences—This is the point where the event that has occurred is no longer acceptable and an investigation is launched.
5. Deepest Significant Underlying Causes—This means at what point do we stop drilling down and decide to take corrective actions?

This definition certainly encompasses and embodies the intent of the DOE guideline for RCA.

WHY DO UNDESIRABLE OUTCOMES OCCUR? THE BIG PICTURE

We must put aside the industry that we work in and follow along from the standpoint of the human being. In order to understand why undesirable outcomes exist, we must understand the mechanics of failure. Virtually all undesirable outcomes are the result of human errors of omission or commission (or decision errors). Experience in industry indicates that any undesirable outcome will have, on average, a series of 10 to 14 cause-and-effect relationships that queue up in a particular pattern in order for that event to occur.

This dispels the commonly held myth that one error causes the ultimate undesirable outcome. All such undesirable outcomes will have their roots embedded in the physical, human, and latent areas.

Physical Roots: are typically found soon after errors of commission or omission. They are the first physical consequences resulting from a human decision error. Physical roots, as will be described in detail in coming chapters, are in essence tangible.

Human Roots: are decision errors. These are the actions (or inactions) that trigger the physical roots to surface. As mentioned previously, these are the errors of omission or commission of the human being.

Latent Roots: are the organizational systems that are flawed. These are the support systems (i.e., procedures, training, incentive systems, purchasing habits, etc.) that are typically put in place to help our workforce make better decisions. Latent roots are the expressed intent of the human decision-making process.

ARE ALL RCA METHODOLOGIES CREATED EQUALLY?

There are many providers of various RCA methodologies on the market today. Many of these providers use tools that are considered RCA in the RCA community and many do not. Many have been in the RCA business for decades, and many have just gotten into it. The point here is that this is a buyer-beware field.

Companies interested in shopping for "RCA" based solely on initial price should hand out a pencil and piece of paper and just ask their employees to ask themselves "WHY?" five times, and they will have their answers.

For those companies looking to make dramatic strides in their operations, shopping on price alone will not cut it. Those who are serious about RCA being a major contributor to their bottom line will be interested in the methodologies involved and what supporting infrastructure may be required to be successful. We will discuss both of these very important topics in detail in coming chapters.

Many of the most respected providers in the RCA industry normally have their own unique styles and vocabularies, but there are also many commonalities among them. PROACT® is no different. These unique qualities are what make the different "brands" of RCA proprietary to a certain provider. It makes them stand out and separates them from the general commodity term of "RCA."

For the users, this is both good and bad. It is good to have variation and competition in the market to keep investments down and provide choices for specific work environments. It is sometimes bad because no generally accepted "standards" emerge to give users a foundation in which all true RCA methods should comply. Also, because there are so many RCA methods on the market, the use of terminology is, at best, inconsistent when comparing them. This further confuses the users when they try to compare terms like our Physical, Human, and Latent Root Causes with terms like Contributing Factors, Primary Root Causes, Underlying Root Causes, Approximate Root Causes, Near Root Causes, Mitigating Factors, Exacerbating Factors, Proximate Causes, Near Root Causes, etc.

ATTEMPTING TO STANDARDIZE RCA—IS THIS GOOD FOR THE INDUSTRY?

Valiant attempts have been made by the joint provider and user communities to develop such a standard for industry. One such attempt was to model it after the SAE JA-1011 RCM Standard.* Debates arose as to whether such a standard is needed at all, and if so, can one be developed without constraining the task of RCA itself? Because RCA requires such open boundaries to the disciplined thought process required to find the truth, would developing a standard bias possible outcomes?

* Evaluation Criteria for Reliability-Centered Maintenance (RCM) Processes, G-11 Supportability Committee, SAE Standards, Document # JA1011, August, 1999 (http://www.sae.org/servlets/productDetail?PROD_TYP=STD&PROD_CD=JA1011_199908).

Creating an RCA standard may define the boundaries of RCA differently than some providers' methodologies. In some circumstances, some providers' established RCA methodologies may be deemed "noncompliant." This would obviously be a detriment to their businesses and naturally they would oppose the development of such a standard. For instance, if an RCA standard listed the validation of each hypothesis with hard evidence as "essential" to RCA, then typical brainstorming techniques would be noncompliant. If another essential element were that the team members had to create the logic by exploring the possibilities of how something could have occurred, then the use of "pick-list" RCA methodologies would be noncompliant.

"Pick-list" RCA is where the methodologies either provide paper templates with their list of possibilities or, if they are software oriented, drop-down lists appear with the vendor's possibilities provided. While these approaches on the surface seem to be the logical and easiest route, there are dangers. One such danger is that the user believes that *all* the possibilities that could have contributed to the undesirable outcome are provided in this list. That will likely never be the case as no vendor can claim to capture all of the variables associated with any event in every environment. The second danger, and perhaps the greatest, is that the task of RCA is meant to raise the knowledge and skill levels of the workforce. A methodology that provides what appears to be all of the answers does not force the users to explore the possibilities on their own, and therefore they do not learn. They are simply doing "paint-by-the-numbers" RCA.

Unfortunately, for the user community especially, this endeavor to develop a common standard never came to pass because the major providers could never come to consensus (which is not unusual). If readers wanted to take it upon themselves, on behalf of their corporations, to develop an RCA standard internally that outlines the essential elements of an analysis process in order for it to be considered RCA, we would encourage them to obtain a copy of the SAE JA-1011 RCM Standard and use it as a baseline draft for the development of a similar document for Root Cause Analysis in their organization.

This SAE Standard is not biased to any provider or methodology. It simply clarifies for the organization what it considers to be the *essential elements* of an RCA. This is important because there are divided camps on what the scope of an RCA is. Some feel the tasks of identifying qualified candidates for RCA is not RCA itself. Some feel that the writing of recommendations and their subsequent approval process and implementation is not in the scope of RCA. So having such a document clarifies what the company considers to be RCA and, more importantly, what is not considered RCA.

WHAT IS NOT ROOT CAUSE ANALYSIS?

In order to recognize what is Root Cause Analysis and what is NOT Root Cause Analysis (Shallow Cause Analysis), we would have to define the criteria that must be met in order for a process and its tools to be called Root Cause Analysis. In the absence of a universally accepted standard, let's consider the following essential elements* of a true Root Cause Analysis process:

* Latino, Robert J. PROACT Approach to Healthcare Workshop. January 2005. www.proactforhealthcare.com

1. Identification of the *Real* Problem to be Analyzed in the First Place. About 80% of the time we are asked to assist on an investigation team, the problem presented to us is not the problem at hand.
2. Identification of the Cause-and-Effect Relationships that Combined to Cause the Undesirable Outcome. Being able to correlate deficient systems directly to undesirable outcomes is critical. Using categorical approaches (as we will explain in Chapter 10, "Analyzing the Data") will often yield less comprehensive results than cause-and-effect approaches).
3. Disciplined Data Collection and Preservation of Evidence to Support Cause-and-Effect Relationships. It is safe to say that if we are not collecting data to validate our hypotheses, we are not properly conducting a comprehensive RCA.
4. Identification of All Physical, Human, and Latent Root Causes Associated with Undesirable Outcome. If we are not identifying system deficiencies that lead to poor decision making, then, again, we are not properly conducting a comprehensive RCA.
5. Development of Corrective Actions/Countermeasures to Prevent Same and Similar Problems in the Future. If we have merely developed good recommendations but never implement them, then we will not be successful in our RCA efforts. This is where the ball is often dropped as well-intentioned people are pulled away by reactive work, and these proactive opportunities fall by the wayside.
6. Effective Communication to Others in the Organization of Lessons Learned from Analysis Conclusions. One of the greatest benefits of a successful RCA is the dissemination of the lessons learned in an effort to avoid recurrence elsewhere in the organization. Oftentimes, successful analyses end up in a paper filing system only to be suppressed from those who could benefit from the lessons learned in the analysis.

Given the above, let's review the basics of some common RCA processes (as opposed to tools, which will be described in detail in Chapter 10). We will call them "RCA" processes because that is the perception of the users. However, as we will explain, the reality is that many of these processes do not meet the minimum requirements of a true RCA process.

Troubleshooting is usually a "Band-Aid" type of approach to fixing a situation quickly and restoring the status quo. Typically, troubleshooting is performed by individuals as opposed to teams and requires little if any proof or evidence to back up assumptions. This off-the-cuff process is often referred to as RCA but clearly falls short of the criteria to qualify as RCA.

Brainstorming is traditionally performed by a collection of experts who throw out a series of disconnected ideas as to the causes of a particular event. Usually such sessions are not structured in a manner that explores cause-and-effect relationships. Rather, people just express their opinions and come to a consensus on solutions. When comparing this approach to the essential elements listed above, brainstorming falls short of the criteria to be called RCA and therefore falls into the Shallow Cause Analysis category.

Analytical Process	Disciplined Data Collection Required?	Typically Team (T) Versus Individual (I) Based	Formal Cause and Effect Structure	Requires Validation of Hypotheses Using Evidence	Identification of Physical (P), Human (H), and Latent (Latent) Root Causes
Brainstorming	N	T	N	N	P or H
Troubleshooting	N	I	N	N	P
Problem Solving	N	T	N	N	P or H
Root Cause Analysis	Y	T	Y	Y	P, H, and L

FIGURE 2.1 Comparison of analytical processes to RCA essential elements.

Problem solving comes the closest to meeting the RCA criteria. Problem solving usually is team based, plus it employs the use of a structured tool(s). Some of these tools may be cause-and-effect based; some may not be. Problem solving oftentimes falls short of the RCA criteria because it does not require evidence to back up what the team members hypothesize. When assumption is permitted to fly as fact in a process, it is not RCA. Figure 2.1 is a quick-reference guide to what the norms are for these types of analytical approaches.

The acceptance of common brainstorming techniques such as the Fishbone Diagram, the 5 Why's, and Process Flow Mapping Techniques have provided many a false sense of security. This false sense of security comes from the belief that these techniques are comparable to the same standard as RCA. Again, this reinforces the need for an internal standard that defines the minimum *essential elements* to be considered RCA in the organization.

The aforementioned processes are not considered RCA processes within the RCA community. This is because they are not typically based on fact. They typically allow ignorance and assumption (hearsay) to be viewed as fact. These are attractive techniques to such a reactive environment because they can be concluded very quickly, oftentimes in a single session, with minimal participation (if any).

Why do such techniques conclude so quickly? They conclude quickly because time usually is not required to collect data or evidence to support the hearsay (hypotheses). Usually, data collection and testing requires the bulk of the time in any investigative occupation. In accident investigations, think of what weight they would carry without providing hard evidence. If the National Transportation Safety Board (NTSB) didn't collect evidence at airline crash scenes, what credibility would they have when issuing conclusions and recommendations? What weight would a prosecutor's case in court carry if the prosecutor had no evidence except hearsay?

HOW TO COMPARE DIFFERENT RCA METHODOLOGIES WHEN RESEARCHING THEM

When researching RCA methodologies, we should consider characteristics other than investment. While the initial investment may be very inexpensive, our greatest concerns should be that the methodology has the breadth and depth to uncover *all* of the root causes associated with any undesirable outcomes. If we focus on cost and not value, we may find that the lifecycle costs to support an inexpensive RCA methodology will be 100 times the original investment when the undesirable outcomes continue to persist and upset the daily operations.

We suggest that when a facility has properly researched the various RCA methodologies on the market, they short-list the top three providers based on their company's internal requirements (i.e., like the standard that we discussed previously). It is also advised that the short-listed providers submit references prior to any future meetings with them. Discussions with these references should focus on comprehensiveness of approach, efficiency and effectiveness, necessary management support, and general acceptance by organizational personnel. We would be seeking to sift out the advantages and disadvantages of the providers' approaches that these users have experienced. We want to be sure to understand issues that are under the control of the provider and issues that are under the control of the purchasing organization. For instance, an organization may select the best RCA option for its environment, but if the management support infrastructure is not in place and the effort fails, it may not be due to a flaw in the selected methodology.

Once short-listed the providers should be given the opportunity to present their approaches either in person or via live on-line conferencing technologies. This is where they should be questioned and evaluated based on the merits of their approaches and the breadth and depth of their offerings. Keep in mind that this will also require preparation on the analyst's side in terms of preparing educated and detailed questions related to the methodology and not just pricing structure.

One tool we provide prospects that are researching RCA methodologies is the evaluation tool shown in Figure 2.2. This is an unbiased way of equally evaluating several approaches based on custom weighting of methodology characteristics.

Notice the characteristics (in this case) in which we have decided to compare the methods short-listed:

1. *Simplicity/User Friendliness*—One thing to all of us that is an endangered species is *time*. Therefore, when conducting such analyses, the methodology must be very simple to grasp in concept and execute in practice.
2. *Analysis Flexibility*—Too much rigidity in a methodology can impose unrealistic constraints that can stifle the analysis itself. As we tell our clients, we are consultants and we live in this ideal world where we make things look so simple. The fact is the best we can do is provide an ideal framework for conducting RCA. The methodology must be pliable enough to work effectively when molded to meet the reality of the working environment.
3. *Initial Cost*—While this is an important characteristic due to our budgeting constraints, we must not let initial cost cloud lifecycle costs and value. If we always opt for the least expensive option, we must consider that if the methodology is inferior and the problem happens again, how much did the RCA purchase really cost the organization?
4. *Quality of Materials*—When the providers are gone, how good is the reference material that you will rely on in their absence?

1. Company X
2. Company Y
3. Company Z

	Evaluation Criteria							
Vendor	**Simplicity/User Friendliness**	**Analysis Flexibility**	**Initial Cost**	**Quality of Materials**	**Results and Reports**	**Training Flexibility**	**Process Credibility and Thoroughness**	**Ability to Track Bottom-Line Results**
1. Company X	5	5	1	5	5	5	3	5
2. Company Y	3	3	5	4	3	5	5	1
3. Company Z	1	3	5	4	3	5	5	1
Weight of criteria	4	3	5	2	4	2	5	3

Note: Process criteria and their alternative ranking are displayed in the table above, which was used to determine the preferred choice. Rankings are based on 5 = Best. The weight of each criteria shown below the table is based on 5 = Most Important.

Final Scores	
Company X	110
Company Y	104
Company Z	96

FIGURE 2.2 Vendor evaluation tool.

5. *Results and Reports*—How well does the approach's reporting capability allow me to meet my compliance obligations and reporting to my superiors? Does the methodology provide me a means for making the business case for implementing my recommendations? What feedback did we receive from the references regarding the reality of results?
6. *Training Flexibility*—Is the training extensive enough that my analysts will be comfortable in doing analyses when the consultant leaves? Will the training involve canned examples in my industry and/or the use of current problems in my facility? Does the training convey knowledge (lecture) and skill (exercises)? Is there follow-up or refresher training available and/or included? Will upper management be trained in an overview format in what their responsibilities will be to support the RCA effort?
7. *Process Credibility and Thoroughness*—What attributes does this approach have that will allow it to likely capture issues that other approaches will not? How easy will it be for my people to bypass the discipline of the RCA process resulting in shortcuts that can increase the risk of recurrence of the undesirable outcome?
8. *Ability to Track Bottom-Line Results*—Does this methodology put any emphasis on Return on Investment (ROI)? What training and tools are provided to ensure that the analysts are capable of making a business case for their analysis results?

Remember, these are only a sampling of criteria in which RCA methodologies can be evaluated. The organization's evaluation team should come up with its own list based on its own needs. Once the criterion has been established, the evaluation team can weight each of these factors as to its importance to the overall decision. We typically use a weighting scale of 1 to 5 where "1" has the lowest impact on the decision and "5" has the greatest. Once these are established and entered into a simple spreadsheet like in Figure 2.2 (after the evaluation team meets with each provider), team members will fill out this evaluation form individually and then average them together as a team.

When the individual forms are compared, if there are great disparities in any particular criteria it should be a signal that further discussion is needed to understand why there is such a gap in how team members view the same thing. This approach is a quick and unbiased way to compare offerings of any kind, not just RCA.

WHAT ARE THE PRIMARY DIFFERENCES BETWEEN SIX SIGMA AND RCA?

Where does RCA fit in Six Sigma? The focal point of most any Six Sigma effort will be to achieve precision through the minimization of process variation. However, the goal of RCA is not to minimize process variation, but to eliminate the risk of recurrence of the event that is causing the variation.

For instance, if a bottling operation was the system being analyzed, Six Sigma might seek to minimize the consequences of "line jams" (process variation) by implementing recommendations that would catch any jams at an earlier state in order to fix them and minimize the production consequences (MTTR—Mean-Time-to-Repair [or Restore]). RCA, on the other hand, would seek to drill down on the individual types of identified line jams and understand the chain of events that led to the jam in the first place. RCA would uncover the system deficiencies that triggered poor decisions being made that set off a series of physical consequences until the line production was affected. *RCA seeks to understand what causes the undesirable outcomes to occur, and Six Sigma seeks to minimize the consequences of those events when they do occur (i.e., process variation).*

Traditionally, Six Sigma toolboxes utilize many Total Productive Maintenance/Management (TPM) problem-solving, brainstorming, and RCA tools such as 5 Why's, Fishbone Diagrams, Fault Tree Analysis, and Timeline Analysis. While these tools are good for basic problem solving, they are not traditionally used to the extent that Root Cause Analysis will be described in this text. RCA tools used in Six Sigma tend to fall short of the depth achieved in real RCA. Oftentimes this lack of depth results in the use of the term "shallow cause analysis."

Once an organization has identified what its RCA needs are, it must then understand the social ramifications of implementing such behavioral changes. Remember, RCA is a thought process and not a tangible product. It involves the complexity and variability associated with the human mind. It involves cultural considerations. Before we delve deeply into the management systems required to support such an effort, we will explore the reasons such efforts often fail. Again, we will learn from those in the past who have paved the way for us.

OBSTACLES TO LEARNING FROM THINGS THAT GO WRONG

In an informal on-line poll* presented to a group of beginner and veteran RCA practitioners, the following question was asked on the RCA discussion forum:

> What are the obstacles to learning from things that go wrong?

The following list is a summary of the responses grouped into appropriate categories by the moderator. Some examples of the actual responses are listed below each category to help further define what was meant by the category title.

1. RCA is almost contrary to human nature—28%
   a. People don't like to admit they made the mistake.
   b. Accountability. If you are the boss—that is it!
   c. We are unwilling to change our own behavior.
2. Incentives and/or priority to do RCAs are lacking—19%
   a. It is not expected of them.
   b. There is no personal incentive to do so.
   c. The work environment does not condone, or accommodate, such a proactive activity.

* Nelms, Robert (2004). What Are the Obstacles to Learning from Things That Go Wrong? [Online]. Available at http://www.rootcauselive.com

3. RCA takes time/we have no time—14%
   a. People are too busy due to daily work/problems.
   b. Variations on "I'm too busy."
4. Ill- or misdefined RCA processes—12%
   a. No agreement on either "how far back" you have to go in your analysis.
   b. Vaguely defined processes.
   c. It is a theoretical approach. It is practically impossible.
5. Our "Western Culture"—9%
   a. The stock market—short-term focus.
   b. Managers being rewarded for short-term results.
   c. The tyranny of the urgent.
6. We haven't had to do RCA in the past—why now—8%
   a. Not how I was trained, not how I/we do things.
   b. Some behavior is so entrenched that it would be like being struck by lightning for some individuals to be aware of the need.
7. Most people don't understand how important it is to learn from things that go wrong—5%
   a. It never occurs to most people that learning from experience is a cost-effective activity.
8. RCAs are not my responsibility—5%.
   a. It's NIMBY (not in my backyard).
   b. That's not our job.

This poll is cited to make an extremely important point to executives. As you can see from the list, every single objection is the result of an improper, inadequate, or nonexistent management support structure. Every one of these objections can be overcome with proper strategy, development, and implementation of a support structure. As a matter of fact, few of these are even related to methodology considerations.

Conversely, not addressing the support structure will likely make such proactive efforts a lip-service effort that is not capable of producing substantial results. An organization can have the best analysts and the best tools, but without proper support the proactive efforts are not likely to succeed.

The following chapter is a training model developed by Reliability Center, Inc. (RCI)* to provide guidance for the design and implementation of a support infrastructure for proactive activities such as RCA. It encompasses not only the elements about specific training objectives necessary to be successful, but also outlines the specific requirements of the executives/management, the Champions, and the Drivers who are accountable for creating the environment for RCA to be successful.

Specific information will be outlined from this model that is pertinent to creating the environment for RCA to succeed. For the sake of this text, we will focus on RCA being the primary proactive activity to support; however, the reader should recognize that the model will fit any proactive initiative.

---

* Reliability Center, Inc. (2004). The Reliability Performance Process (TRPP). Hopewell: Reliability Center, Inc.

3 Creating the Environment for RCA to Succeed

The Reliability Performance Process (TRPP®)

TRPP* is an RCA management support model developed by Reliability Center, Inc. (RCI). Not only does it encompass the elements of specific training objectives necessary to be successful, but it also outlines the specific requirements of the executives/management, the Champions, and the Drivers who are accountable for creating the environment for RCA to be successful.

We will be outlining specific information from TRPP that is pertinent to creating an environment for RCA to succeed.

THE ROLE OF EXECUTIVE MANAGEMENT IN RCA

As with implementing any initiative into an organization, the path of least resistance is typically from the top down, relative to the bottom-up approach. The one thing we should always be cognizant of is the fact that no matter what the new initiative is, the end user will likely view it as the "program-of-the-month." This should always be in the back of your mind when developing implementation strategies.

Our experience is the closer we get to the field where the work is actually performed, the sharp end, the more skeptics we encounter. Every year a new organizational "buzz" fad emerges and the executives hear and read about it in trade journals, magazines, and business texts. Eventually, directives are given to implement these "fads," and by the time they reach the sharp end, the well-intentioned objectives of the initiatives are so diluted from miscommunication that they are viewed as non-value-added work and a burden to an existing workload. This is the paradigm of the end user that must be overcome for successful implementation of RCA.

Oftentimes when we look at institutionalizing these types of initiatives, we look at them strictly from the shareholders' view and work backward. Don't get us wrong—we are not against new initiatives that are designed to change behavior for the betterment of the corporation. This process is necessary to progress as a society. However, the manner in which we try to attain that end is what has been typically ineffective.

Changing behavior is essentially changing culture, and changing culture takes time. You must take into consideration the shared values of each site and link the

* TRPP is a registered trademark of Reliability Center, Inc.

new information so that it fits the values of those involved. To create behavior change the new information must become part of the belief system of the individuals who are expected to change.

We must look at linking what is different about the initiative from the perception of the end user in comparison to others that have not succeeded. We must look at the reality of the environment of the people who will make the change happen. How can we change the behavior of a given population to reflect those behaviors that are necessary to meet our objectives?

Let's take an example. If I am a maintenance person in an organization and have been so for my entire career, I am expected to repair equipment so that we can make more product(s). As a matter of fact, my performance is measured by how well I can make the repair in the shortest time frame possible. I am given recognition when emergencies occur, and I respond almost heroically. This same scenario can apply to the service industry, healthcare, and anywhere else where people spend most of their days reacting to problems as opposed to working on opportunities.

Now along comes this Root Cause Analysis (RCA) initiative and they want me to participate in making sure that failures do not occur anymore. In my mind, if this objective is accomplished, I am out of a job! Rather than be perceived as NOT being a team player, I will superficially participate until the "program-of-the-month" has lived out its average 6-month shelf life and then go on with business as usual. We have seen this scenario repeatedly, and it is a valid concern based on the reality of the end user. This perception must be overcome prior to implementing an RCA initiative in an organization.

We must face the fact that we are in a global environment. We must compete not only domestically, but with foreign markets as well. Oftentimes these markets have an edge in that their costs to produce are significantly less than those in the United States. Maintenance, in its true state, is often viewed as a necessary evil to a corporation. But when equipment fails, it generally holds up production, which holds up delivery, which holds up profitability. Imagine a world where the only failures that occurred were wear-out failures that were predictable. This is a world we are moving toward, as precision environments increasingly become the expectation. As we move in this direction, there will be less need for maintenance-type skills on a routine basis.

What about the area of Reliability Engineering (RE)? Most organizations we deal with never have the resources to properly staff their Reliability Engineering groups. There are plenty of available roles in the field of Reliability. Think about how many Reliability jobs are available: vibration analysts, root cause analysts, infrared thermographers, metallurgists, designers, inspectors, nondestructive testing specialists, and many more.

We are continually intrigued by the most frequently used objection to using RCA at the sharp end: "I don't have time to do RCA." If you think hard about this statement, it really is an oxymoron. Why do people typically NOT have time to do RCA? They are so busy fire fighting that they do not have time to analyze why the undesirable outcome occurred in the first place. If this remains as a maintenance strategy, then the organization will never progress, because no dedication is being put toward "getting rid of the need to do the reactive work!"

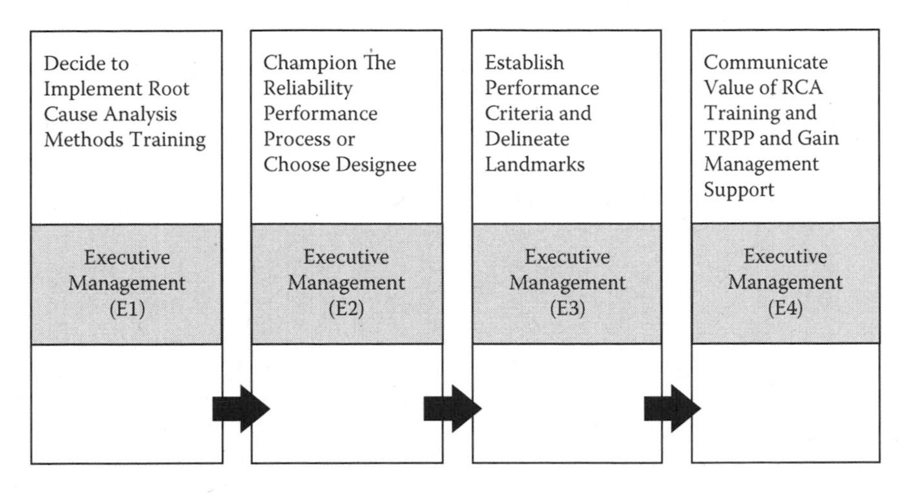

FIGURE 3.1 TRPP executive management roles.

So how can executives get these very same people to willingly participate in a new RCA initiative (Figure 3.1)?

1. It must start with an executive putting a rubber stamp on the RCA effort and outlining specifically what his or her expectations are for the process and a time line for when he or she expects to see bottom-line results.
2. The approving executive(s) should be educated in the RCA process themselves, even if it is an overview version. Such demonstrations of support are worth their weight in gold because the users can be assured that the executives have learned what the users are learning and agree and support the process.
3. The executive responsible for the success of the effort should designate a Champion or Sponsor of the RCA effort. This individual's roles will be outlined later in this chapter.
4. It should be clearly delineated how this RCA effort will benefit the company, but more importantly, it should also delineate how it will benefit the work life of every employee and provide a quality product for the customers.
5. Next, the executive should outline how the RCA process will be implemented to accomplish the objectives and how management will support those actions.
6. A policy or procedure should be developed to institutionalize the RCA process. This is another physical demonstration of support that also provides continuity of the RCA application and perceived staying power. It gives the effort perceived staying power because even if there is a turnover in management, institutionalized processes have a greater chance of weathering the storm.
7. However, the most important action an executive can take to demonstrate support is to sign a "fat" check. We believe this is a universal sign

of support. Any organization who has implemented SAP®* or Six Sigma should be familiar with this concept.

8. The executive management should craft an incentive program that will insure that the Champions will be well compensated for success. This can mean promotional opportunities and/or bonuses. Typically about 5% of the savings from the Reliability initiative will be enough to compensate the Champions appropriately and deliver a substantial payback to the company.

THE ROLE OF AN RCA CHAMPION (SPONSOR)

Even if all these actions take place, it does not automatically insure success.

How many times have we all seen a well-intentioned effort from the top try to make its way to the field and fail miserably? Typically, somewhere in the middle of the organization, the translation of the original message begins to deviate from its intended path. This is a common reason why some very good efforts fail—because of the miscommunication of the original message!

Because of this breakdown there must be a Corporate Champion role and Site Champion role. The Corporate Champion's role is significant because this person is responsible for developing Champion criteria and selecting the Site Champions. The Corporate Champion is responsible for developing the metrics and getting the metrics approved by the corporate oversight committee. The approved metrics are the scorecard used to measure RCA success. The Corporate Champion is responsible for approving the implementation plan created by the Site Champions, as well as serving as the common link between the site and the corporate management. The Corporate Champion must keep all of those concerned abreast of the progress as well as any barriers hindering the sites from successful implementation. This means there will be quarterly meetings with the Corporate Oversight Committee as well as bimonthly meetings with Site Champions.

If we are proactive in our thinking and we foresee a barrier to success, then we can plan for its occurrence and avoid it. This is where the role of the RCA Site Champion comes into play. We will use the term "Champion" synonymously with the term "Sponsor."

There are three major roles of a Site RCA Champion (Figure 3.2):

1. The Site Champion must administer and support the RCA effort from a management standpoint. This includes ensuring that the message from the top to the floor is communicated properly and effectively. Any deviations from the plan will be the responsibility of the Site Champion to align or get back on track. This person is truly the "Champion" of the RCA effort.
2. The second primary role of the RCA Site Champion is to be a mentor to the Drivers and the Analysts. This means the Champion must be educated in the RCA process and have a thorough understanding of what is necessary for success.

* SAP is a registered trademark of SAP AG.

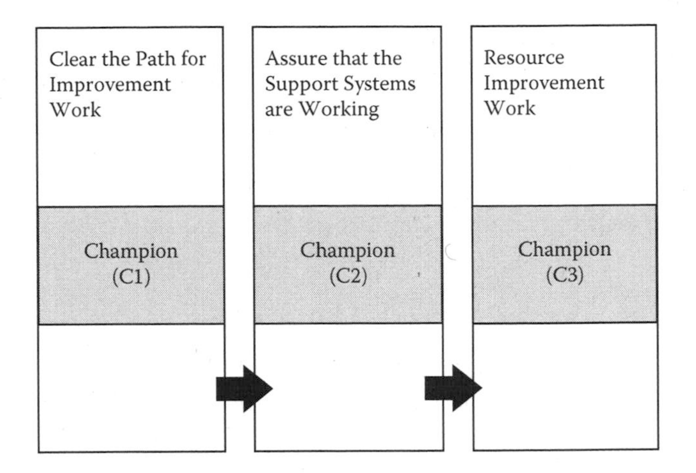

FIGURE 3.2 TRPP Champion roles.

3. The third primary role of the RCA Site Champion is to be a protector of those utilizing the process and uncover causes that may be politically sensitive. Sometimes we refer to this role as providing "air cover" for ground troops. In order to fulfill this responsibility, the RCA Site Champion must be in a position of authority to take a defensive position and protect the person who uncovered these facts.

Ideally, this is a full-time position. However, in reality, we find it typically to be a part-time effort for an individual. In either situation we have seen it work; the key is it must be made a priority to the organization. This is generally accomplished if the executive(s) perform their designed tasks set out above. Actions do speak louder than words. When new initiatives come down the pike and the workforce sees no support, it becomes another "they are not going to walk-the-talk" issue. The initiatives are viewed as lip-service programs that will pass over time. If the RCA effort is going to succeed, it must first break down the paradigms that currently exist. It must be viewed as different from the other programs. It is also the RCA Site Champion's role to project an image that this is different and will work.

The RCA Champion's additional responsibilities include insuring that the following responsibilities are carried out (Figure 3.3):

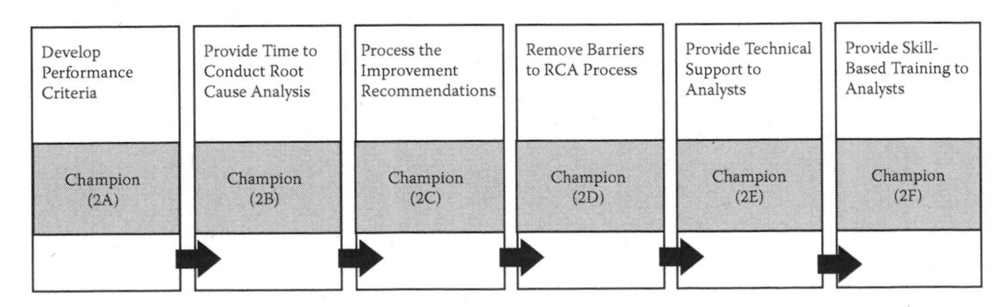

FIGURE 3.3 TRPP additional roles of Champions/management.

1. Selecting and training RCA Drivers who will lead RCA teams. What are the personal characteristics that are required to make this a success? What kind of training do they need to provide them the tools to do the job right?
2. Developing management support systems such as
   a. *RCA performance criteria*—What financial returns are expected by the corporation? What are the time frames? What are the landmarks?
   b. *Providing time*—In an era of reengineering and lean manufacturing, "How are we going to mandate that designated employees WILL spend 10% of their week on RCA teams?"
   c. *Process the recommendations*—How are recommendations from RCAs going to be handled in the current work order system? How does improvement (proactive) work get executed in a reactive work order system?
   d. *Provide technical resources*—What technical resources are going to be made available to the analysts to prove and disprove their hypotheses using the "Whatever It Takes" mentality?
   e. *Provide skill-based training*—How will we educate RCA team members and ensure that they are competent enough to participate on such a team?
3. The Site Champion will also be responsible for setting performance expectations. The Site Champion should draft a letter that will be forwarded to all students attending the RCA training. The letter should clearly outline exactly what is expected of them and how the follow-up system will be implemented.
4. The Site Champion should ensure that all training classes are kicked off either by the Site Champion him- or herself, an executive, or another person of authority, giving credibility and priority to the effort.
5. The Site Champion should also be responsible for developing and setting up a recognition system for RCA successes. Recognition can range from a letter by an executive to tickets to a ball game. Whatever the incentive is, it should be of value to the recipient.

Needless to say, the role of a Champion is critical to the RCA process. The lack of a Champion is usually why most formal RCA efforts fail. There is no one leading the cause or carrying the RCA flag. Make no bones about it; if an organization has never had a formal RCA effort, or it had one that failed, such an endeavor is an uphill battle. Without an RCA Champion, sometimes it can feel like you are on an island by yourself.

THE ROLE OF THE RCA DRIVER

The RCA Driver can be synonymous with the RCA Team Leaders. These are the people who organize all the details and are closest to the work. Drivers carry the burden of producing bottom-line results for the RCA effort. Their teams will meet, analyze, hypothesize, verify, and draw factual conclusions as to why undesirable outcomes occur. Then they will develop recommendations or countermeasures to eliminate the risk of recurrence of the event.

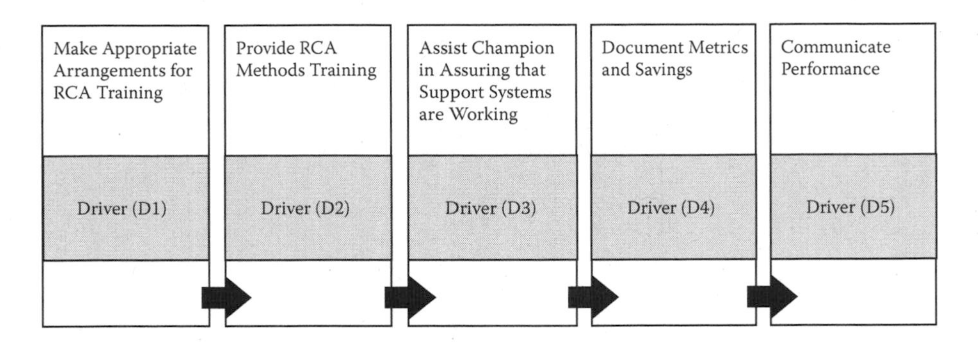

FIGURE 3.4 TRPP Driver roles.

All the executives', managers', and Champions' efforts to support RCA are directed at supporting the Driver's role to ensure success. The Driver is in a unique position in that he or she deals directly with the field experts—the people who will comprise the core team. The personality traits that are most effective in this role as well as the role of a core team member will be discussed at length in Chapter 9.

From a functional standpoint, the RCA Driver's roles are (Figure 3.4)

1. *Making arrangements for RCA training for team leaders and team members*—This includes setting up meeting times, approving training objectives, and providing adequate training rooms.
2. *Reiterating expectations to students*—Clarify to students what is expected of them, when it is expected, and how it will be obtained. The Driver should occasionally set and hold RCA class reunions. This reunion should be announced at the initial training so as to set an expectation of demonstrable performance by that time.
3. *Ensure that RCA support systems are working*—Notify the RCA Champion of any deficiencies in support systems and see that they are corrected.
4. *Facilitate RCA teams*—The Driver will lead the RCA teams and be responsible and accountable for the teams' performance. The Driver will be responsible for properly documenting every phase of the analysis.
5. *Document performance*—The Driver will be responsible for developing the appropriate metrics against which to measure performance. This performance will always be converted from units to dollars when demonstrating savings and thus success.
6. *Ensure regulatory compliance*—The Driver will be responsible for ensuring that the analyses conducted are thorough and credible enough to meet applicable regulatory standards and guidelines.
7. *Communicate performance*—The Driver will be the chief spokesperson for the team. The Driver will present updates to management as well as to other individuals on site and at other similar operations that could benefit from the information. The Driver will develop proper information distribution routes so that the RCA results get to others in the organization who may have, or have had, similar occurrences.

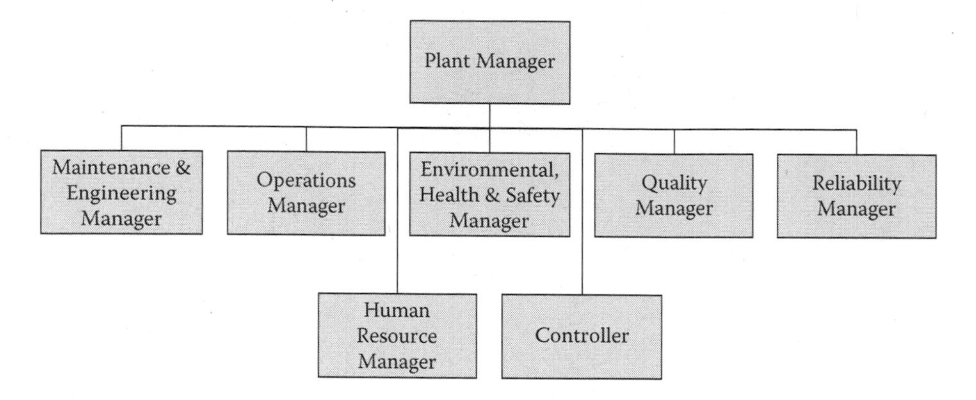

FIGURE 3.5 Ideal position for Reliability on organizational chart.

The Driver is the last of the support mechanisms that should be in place to support an RCA effort. Most RCA efforts that we have encountered are put together at the last minute as a result of an "incident" that just occurred. We discussed this topic earlier regarding using RCA as only a reactive tool.

A structured RCA effort should be properly placed in an organizational chart (Figure 3.5). Because RCA is intended to be a proactive task, it should reside within the control of a structured Reliability Department. In the absence of such a department, it should report to a staff position such as a VP of Operations, VP of Engineering, VP of Quality, and/or VP of Risk. Whatever the case may be, ensure that the RCA effort is never placed under the control of a Maintenance Department (or any other reactive department). By its nature, a Maintenance Department is a reactive entity. Its members' role is to respond to the day-to-day activities in the field. The role of a true Reliability Department is to look at tomorrow, not today. Any proactive task assigned to a Maintenance Department is typically doomed from the start.

This is why, when "Reliability" became the buzzword of the mid 1990s, many Maintenance Engineering Departments were renamed as Reliability Departments. The same people resided in the department and they were performing the same jobs; however, their title was changed and not their function. If you are an individual who is charged with the responsibility of responding to daily problems and also seizing future opportunities, you are likely to never get to realize those opportunities. Reaction wins every time in this scenario.

Now let's assume at this point that we have developed all the necessary systems and personnel to support an RCA effort. How do we know what opportunities to work on first? Working on the wrong events can be counterproductive and yield poor results. In the next chapter we will discuss a technique to use to sell working on one event versus another.

SETTING FINANCIAL EXPECTATIONS: THE REALITY OF THE RETURN

As discussed earlier, one of the roles of the Corporate Champion is to delineate financial expectations of the RCA effort. This will obviously vary with the Key

Performance Indicators (KPIs) of each firm, but in this section we will look at providing a typical business case to justify implementing an RCA effort.

Because the costs to implement such an effort will vary based on each facility, its product sales margin, its labor costs, and the training costs (in-house versus contract), we will base our justifications on the following assumptions:

1. Assumptions
   a. Loaded cost of hourly employee: US$50,000/year
   b. Hourly employees will spend 10% of their time on RCA teams.
   c. Loaded cost of full-time RCA Driver (salaried): US$70,000/year
   d. RCA Driver will be a full-time position.
   e. RCA training costs (hourly): US$400/person per day
   f. RCA training costs (salaried): US$500/person per day
   g. Population trained: Per 100 trained.
2. RCA Return Expectations
   a. Train 100 hourly employees in RCA methods.
   b. Train one salaried employee to lead RCA effort.
   c. Critical mass (assumption): 30% of those trained will actually use the RCA method in the field. This results in 30 personnel trained in RCA methods actually applying it in the field (100 trained × 30% applying).
   d. Of the 30 personnel applying the RCA method, let's assume they are working in teams of three at a minimum. This results in 10 RCA teams applying the methodology in the field (30 personnel/3 per team).
   e. Each RCA team will complete one analysis every 2 months. This results in 60 completed analyses per year (10 RCA teams × 6 analyses/year)
   f. Each "Significant Few" (to be discussed in Chapter 5) analysis will net a minimum of US$50,000 annually. This results in an annual return of US$3,000,000 per 100 people trained in RCA methods.
3. The costs of implementing RCA
   a. Year 1
      i. Training 100 hourly employees in 3 days of RCA: US$120,000
      ii. Training one salaried person in 5 days of RCA: US$2,500
      iii. 10% of 30 hourly employees' time per week, annually: US$150,000
      iv. Salary of RCA Driver/year: US$70,000
      v. Total RCA implementation costs for Year 1: US$342,500
   b. Year 2
      i. Training 100 hourly employees in 3 days of RCA: US$0
      ii. Training one salaried person in 5 days of RCA: US$0
      iii. 10% of 30 hourly employees' time per week, annually: US$150,000
      iv. Salary of RCA Driver/year: US$70,000
      v. Total RCA implementation costs for Year 1: US$220,000*

* All costs of resources to prove hypotheses and implement recommendations are considered as sunk costs. Technical resources are currently available and budgeted for, regardless of RCA. Also, recommendations from RCA generally result

in the implementation of organizational system corrections, for instance, rewriting procedures, providing training, upgrading testing tools, restructuring incentives, etc. These types of recommendations are not generally considered as capital costs. Capital costs resulting from RCA, in our experience, are not the norm but the exception.

4. Return-on-investment
   a. Total expected return—Year 1: US$1,500,000*
   b. Total expected costs—Year 1: US$342,500
   c. ROI Year 1: 437%

* Assumes that it will take 6 months to train all involved and get up to speed with actually implementing RCA and the associated recommendations. This is the reason for cutting this expectation in half for the first year.

   a. Total expected return—Year 2: US$3,000,000
   b. Total expected costs—Year 2: US$220,000
   c. ROI Year 2: 1360%

As you can tell from these numbers, the opportunities are left to the imagination. They are real, and they are phenomenal to the point that they are unbelievable. When we review the process we just went through, look at the conservativeness built in:

1. Only 30% of those trained will actually apply the RCA method.
2. Students will only spend 10% of their time on RCA.
3. Students will work in teams of three or more.
4. Students will only complete one RCA every 2 months.
5. Each event will only net US$50,000/year.

Use this same cost-benefit thought process and plug in your own numbers to see if the ROIs are any less impressive. Using the most conservative stance, it would appear irrational NOT to perform RCA in the field. How many of our engineering projects would be turned down if we demonstrated to management an ROI ranging from 437% to 1360%? Not many!

INSTITUTIONALIZING ROOT CAUSE ANALYSIS (RCA) IN THE SYSTEM

In an era where most college graduates will likely be employed by a minimum of five employers (on average) in their career, stability of turnover is difficult to control. This poses a problem with what is often called "corporate memory." Corporate memory is the ability to retain the knowledge and experience of the workforce in the midst of a high turnover environment. How does a company expect to produce a quality product in a consistent manner when their workforce is inconsistent? This is an especially difficult problem today as the "baby boomer" generation forges through to retirement. When the mass exodus of knowledge and experience occurs in industry, how will we compensate and be able to compete in our global economy?

RCA actually can play a major role in filling this corporate memory void. RCA is a tool that maps out a thought process used to successfully solve a problem. This map in essence is an aggregated thought based on the collective knowledge and experience of our workforce. What we need to do is (1) encourage the activity of RCA in a disciplined manner and (2) electronically catalogue these analyses in a manner so that future employees can view how the past analysts derived their conclusions.

Activity (1) can be accomplished by writing a procedure for RCA that will survive the absence of a previous RCA Champion. The RCA Champion may have moved on to bigger and better things, but what we want to occur is that the activity of RCA is still expected by the organization via policy and/or procedure.

The next section contains a sample RCA procedure* that we have used in industry in the past. It should be used as a draft to model a more accommodating one for an individual facility.

SAMPLE PROACT RCA PROCEDURE (RELIABILITY CENTER, INC.)

1. PURPOSE
   1.1 To provide consistency to the organization in the application of the PROACT Root Cause Analysis (RCA) Process.
   1.2 To provide guidance in the following areas:
   Requests
   Analyses
   Reporting
   Presenting
   Tracking

2. APPLICATION/SCOPE
   This procedure applies to all users of the PROACT process conducted in compliance with all Safety Policies and Procedures unless otherwise directed by the Department Manager.

3. RESPONSIBILITY
   3.1 The Supervisor of Reliability Engineering (or equivalent) shall have the responsibility to review, amend, and revise this procedure as necessary to ensure its integrity and application.
   3.2 The Supervisor of Reliability Engineering (or equivalent) shall have the responsibility to develop, implement, review, and revise related procedures and/or documents required in this procedure.

4. DEFINITIONS
   4.1 **Champion**: usually a person in authority who sponsors and mentors the Principal Analysts and supports the RCA effort.
   4.2 **Charter**: defines the charter (or mission) of the RCA effort.
   4.3 **Chronic Events**: events that occur repetitiously.

* © 1997 Reliability Center, Inc., Sample PROACT RCA Procedure.

4.4 **Critical Success Factor (CSF)**: Identifiable marker that will signal the RCA effort has been successful. Guidelines in which the RCA team operates.
4.5 **Logic Tree**: a graphical representation of logic used to uncover physical, human, and latent root causes.
4.6 **Opportunity Analysis (OA)**: a technique to identify the most important failures (Significant Few) to analyze.
4.7 **Principal Analyst (PA), Qualified**: the individual assigned the responsibility of leading and completing the RCA. The individual is qualified based on his or her successful completion of the PROACT Certification Workshop.
4.8 **PROACT**: a software program that facilitates the PROACT RCA process.
4.9 **Root Cause Analysis (RCA)**: any evidence-driven process that, at a minimum, uncovers underlying truths about past adverse events, thereby exposing opportunities for making lasting improvements.
4.10 **Significant Few**: the 20% of the failure events that have been deemed to be accountable for 80% of the loss. This information is derived from the OA.
4.11 **Sporadic Event**: a one-time catastrophic event.
4.12 **Vital Many**: the many deviations that occur in a facility that equate to continuous improvement efforts.

5. REFERENCES
5.1 Site Policy Manual
5.2 Site Safety Manual
5.3 Site Quality Manual

6. SPORADIC EVENTS
6.1 An RCA is requested for sporadic events with a total cost (maintenance, operations, and lost profit opportunities) greater than $200,000. Listed below are several examples:
Unpredicted Event
Property Damage
Lost Production
6.2 An RCA is requested for incidents that resulted in or could have resulted in personal injury or damage to equipment or property as defined in Section X of the Safe Practices Manual.
6.3 An RCA is requested for repeat customer complaints and complaints from key customers.

7. SIGNIFICANT FEW
A Qualified PA will lead the RCA of the Significant Few events that were identified by the Department OA, unless redirected by the Reliability Coordinator and/or the Department Manager.

7.1 **Assignment of Champion**: The Division Reliability Coordinator will be assigned as the Champion of the event that falls within his or her Division.

7.1.1 A Qualified Principal Analyst (PA) will be assigned as the PA for the Significant Few events assigned to the Department.

8. VITAL MANY/CONTINUOUS IMPROVEMENT

The RCA of the Vital Many events will be led by a PA or other qualified personnel who are not in the Reliability Engineering group.

8.1 **Assignment of Champion**: The Division Reliability Coordinator will be assigned as the Champion of the event that falls within his or her Division.

8.1.1 A PA or other qualified personnel will be assigned or obtained by the Division Reliability Coordinator to lead the RCA.

8.1.2 The Division Reliability Coordinator's role is to provide the resources or obtain the resources that the PA needs to do the job right and to identify and remove obstacles that hinder his or her analysis.

9. DETERMINATION OF TEAM MEMBERS

Certain events will require a team to be formed while others will not. If a team needs to be assembled, the PA will make a recommendation to the Division Reliability Coordinator. The following items also need to be addressed when selecting the team.

- Multidisciplined (i.e., mechanical, electrical, financial, managerial, hourly, etc.)
- Personnel directly affected by problem or event
- Personnel who may be involved with implementation of solution
- Excused from normal work assignments while working on RCA (similar to HAZOP studies)

10. RCA METHODOLOGY

10.1 When a team has been formed that is not familiar with RCA, the team will attend, at a minimum, a 1-day Basic Failure Analysis (BFA) course before proceeding with the analysis.

10.2 The team will accurately define the event.

10.3 The Charter and Critical Success Factors (CSFs) of the analysis need to be developed so each team member knows the purpose of the analysis effort and if the effort is successful.

10.4 Develop Strategy for Collecting the 5P's. The team or PA needs to develop the strategy for capturing the 5P's. This may involve taking pictures, retrieving data from the operating instrumentation, interviewing personnel, etc. The urgency with which this data is collected will depend on whether this is a chronic or sporadic event.

10.5 Assignment of 5P's: The PA will assign the 5P's (listed below) to team members who will be responsible for collecting the data.
- Parts
- Position
- People
- Paradigms
- Paper

10.6 **Analyze**: Using the data collected, develop a logic tree.
10.6.1 The logic tree will not be considered complete unless all the applicable latent roots are identified.

10.7 **Hypothesis Verification**: Each hypothesis block on the logic tree needs to be verified (proven or disproven). This is one of the most crucial steps in the RCA process. Without verification, the findings and recommendations of the RCA are meaningless.

10.8 **RE Review Logic Tree**: The PA will contact the Division Reliability Coordinator when the team is ready to review the logic tree. The review should take place before proceeding with the report and the formal publishing of the analysis in the PROACT Software Program.

10.9 **Write Report**: The report should include the following sections:
- Executive Summary
- Description of Event
- Description of Mechanism
- Review of Causes and Recommendations
- Assignment of Responsibilities and Time Lines
- Detailed/Technical Section
- Detailed Recommendations
- Appendices
- Participants Involved
- 5P's Data Collection Forms
- Verification Logs
- Logic Tree

10.10 **Develop Draft Recommendations:** a presentation of the findings of the RCA shall be given to personnel *affected by* implementing the recommendations and to personnel who *will implement* the recommendations and others as applicable. This will provide input that may affect or change specifics about the recommendations.

10.11 Revise and review the recommendations as necessary.

10.12 Develop corrective action items for each of the recommendations.

10.13 Formally present findings and recommendations to the Reliability Team and/or appropriate management personnel for implementation approval.

11. UTILIZATION OF PROACT RCA SOFTWARE

All documentation of RCAs is to be stored electronically using the PROACT RCA software program on the designated client server. Use of this program shall be in strict accordance with the license to the corporation.

11.1 **User Prerequisites**: All users of PROACT must first successfully complete requisite training in one or more of the following courses based on their participation in the analysis.

11.1.1 **PROACT RCA Methods**: All Principal Analysts (PAs) shall complete the 5-Day RCA Methods course either on site or at a public location. It will be at the discretion of the PA to determine which team members receive the password for password-protected analyses.

11.1.2 **BFA (Basic Failure Analysis)**: All RCA team members shall successfully complete the 1-Day BFA training by a licensed BFA Trainer.

11.1.3 **PROACT Software Training**: All users of PROACT® RCA software shall successfully complete either the 5-Day RCA Methods training or the 1-Day BFA training before becoming eligible for PROACT Software training. All potential PROACT users are required to attend a 4-hour short course in hands-on PROACT instructor-led training.

11.2 The PA shall be responsible for the complete accuracy of the analysis utilizing the software program. Team members shall update their responsibilities in any given analysis; however, the PA is ultimately responsible for reviewing the accuracy and thoroughness of the complete analysis.

11.3 The PA will assume the responsibility for when it is time to publish the RCA. Publishing the analysis in PROACT means that the completed RCA is certified to be credible and thorough. Once published, the analysis serves as a Logic Template for the rest of the corporation. Publishing also means that all sensitive materials have been reviewed by the legal department and have been approved for publishing in this format.

11.4 The PA will reserve the right to password-protect the RCA. Only team members of that specific RCA shall be permitted to have the password. It shall be the responsibility of the PA to remove the password once the RCA has been published.

12. CORRECTIVE ACTION AND TRACKING

Personnel will be assigned responsibility for the corrective actions necessary to implement the recommendations that result from the RCA. These corrective actions will be tracked and a report issued.

12.1 The Division Reliability Coordinator and PA will assign responsibility for the corrective action items unless otherwise directed by the Department Manager or his or her designee.

12.2 The PA will notify a member of the Reliability Group (RG) that the RCA corrective action items have been assigned.

12.3 The PA will see that a copy of the full report (hard copy and electronic) is given to the RG for filing purposes.

12.4 RCAs that result from events that listed Safety Procedures will primarily be handled by Plant Protection or Environmental Affairs. These Departments are responsible for tracking corrective action items that result from these RCAs.

12.5 All RCAs corrective action items (except as noted in 10.5) will be issued as needed in a report to the personnel assigned responsibility for the items. The corrective action items will remain in the report until completed.

12.6 Updates to the report can be forwarded to the Division Reliability Coordinator as they are completed and will be incorporated into the next quarterly report.

12.7 A progress report will be sent to the Department Manager for his or her review.

4 Failure Classification

Before discussing the issue of Root Cause Analysis we must first begin setting the foundation with some key terminology. As mentioned earlier in this text, one of the primary reasons for the misinterpretation of "RCA" is that there is no standard definition against which to benchmark it. Therefore, everyone defines RCA as they please and the result is shallow cause methodologies being equated to root cause methodologies.

For our purposes, let's begin by discussing the key differences between the terms "problems" and "opportunities." There are many people who tend to use these terms interchangeably. However, the truth is that these terms are really at opposite ends of the spectrum.

A problem can be defined as *a negative deviation from a performance norm* (Figure 4.1). What exactly does this mean? It simply means that we cannot perform up to the normal level or standard that we are used to. For example, let's assume we have a widget factory. We are able to produce 1000 widgets per day in our factory. At some point we experience an event that interrupts our ability to make widgets at this level. This means that we have experienced a negative deviation from our performance norm, which in this case is 1000 widgets.

An opportunity is really just the opposite of a problem. It can be defined as *a chance to achieve a goal or an ideal state* (Figure 4.2). This means that we are going to make some changes to increase our performance norm or status quo. Let's look at our widget example again. If our normal output is 1000 widgets per day, then any changes we make to increase our throughput would be considered an opportunity. So if we eliminate certain bottlenecks from the system and start to produce 1100 widgets in a day, this would be considered an opportunity.

Now let's put these terms into perspective. When a problem occurs and we take action to fix it, do we actually improve or progress? The answer to this question is an emphatic NO. When we work on problems, we are essentially working to maintain the status quo or performance norm. This is synonymous with the term "reaction." We react when a problem occurs to get things back to their normal, status quo state. If all we do is work on problems, we will never be able to progress. In our dealings with companies all over the world, we often ask the question, "How much time do you spend reacting versus proacting in your daily routines?" Most surveyed will answer 80% reacting and 20% proacting. If this is true, then there is very little progress being made. This would seem to be a key indicator as to why most productivity increases are minimal from year to year.

Let's consider opportunities for a moment (Figure 4.3). When we work on opportunities do we progress? The answer is YES! When we achieve opportunities, we are striving to raise the status quo to a higher level. Therefore, to progress we have to

FIGURE 4.1 Problem definition graph.

FIGURE 4.2 Opportunity definition graph.

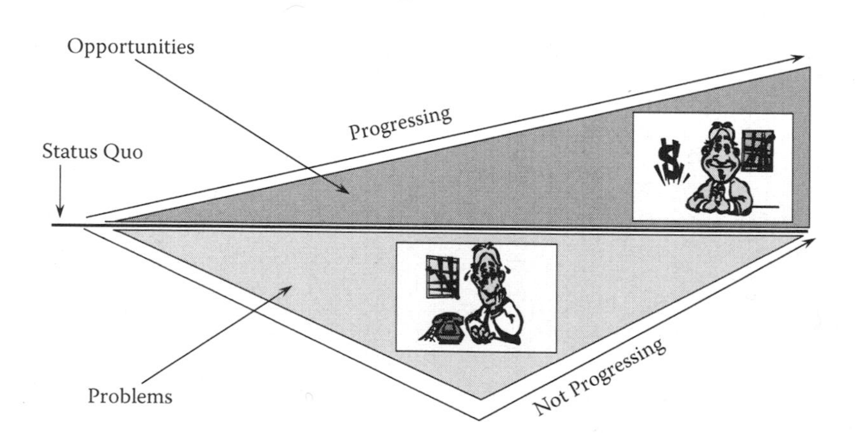

FIGURE 4.3 Opportunity graph.

begin taking advantage of the numerous opportunities presented to us. So if working on problems is like reacting, then working on opportunities is like proacting.

So the answer is simple. We should all start working on opportunities and disregard problems, right? Why can't we do this? There are many reasons, but a few are obvious. Problems are more obvious to us since they take us away from our normal operation. Therefore, they get more attention and priority. We can always put an opportunity off until tomorrow, but problems have to be addressed today. There is also the issue of rewards. People who are good reactors, who come in and save the day, tend to get pats on the back and the old "atta-boys." What a great thing from the reactor's perspective: recognition, overtime pay, and, most importantly, job security. We have seen many cases where the person who tries to prevent a problem or event from occurring gets the cold shoulder while the person who comes in after the event has occurred gets treated like a king. This is not to say that we should not reward exemplary reactors, but we also have to reinforce good proactive behavior as well.

Then there is the risk factor. Which are more risky, problems or opportunities? Opportunities are always more risky since there are many unknowns. With problems there are virtually no unknowns. We usually have fixed the problems before, so we certainly have the confidence to fix them again. I once had a colleague who said, "… when you get really good at fixing something, you are getting way to much practice." In a perfect world, we should have to pull the manual out to see what steps to take to fix the problem. How many times do we a see a craftsman, or even a doctor for that matter, pulling out the manual to troubleshoot a problem? People in this day and age do not want to take a lot of chances with their career, so opportunities begin to look like what we like to call "career limiting" activities. One of the top 10 causes of human error is "overconfidence."

With that said, we have to figure out a way of changing the paradigm that reactive work is always more important than proactive work. This means opportunities are just as important, if not more so, than problems.

Let's switch gears and talk about the different types of failures or events that can occur. Incidentally, when we talk about failures, we are not always talking about machines or equipment. "Failure" can also be unexpected patient deaths, operational upsets, administrative delays, quality defects, or even customer complaints. There are two basic categories of failures that can exist: sporadic and chronic. Let's look at each of these categories in greater detail.

A sporadic (to be used synonymously with acute) occurrence usually indicates that a dramatic event has occurred. For example, maybe we had a fire or an explosion in our manufacturing plant, we lost a long-standing contract to a competitor, or a patient has died unexpectedly. These events tend to demand a lot of attention. Not just attention, but urgent and immediate attention. In other words, everyone in the organization knows something bad has happened. The key characteristic of sporadic events is they happen only once. Sporadic failures have a very dramatic impact when they occur, which is why many people tend to apply financial figures to them. For instance, you might hear someone say, "We had a $10,000,000 failure last year."

Sporadic events are very important and they certainly do cost a lot of money when they occur. The reality, however, is that they do not happen very often. If we had a lot of sporadic events, we certainly would not be in business very long. Sporadic losses can also be distributed over many years. For example, if the engine in your car fails and you need to replace it, it will be a very costly expense, but you can amortize that cost over the remaining life of the car.

Chronic events, on the other hand, are *not* very dramatic when they occur. These types of events happen over and over again. They happen so often that they actually become a cost of doing business. We become so proficient at working on these events that they become part of the status quo. We can produce our "normal" output in spite of these events.

Let's look at some of the characteristics of chronic events. Chronic events are accepted as part of the routine. We accept the fact that they are going to happen. In a manufacturing plant, we will even account for these events by developing a maintenance budget. A maintenance budget is in place to make sure that when routine events occur we have money on hand to fix them. These types of events do, however, demand attention, but usually not the attention a big sporadic or acute event would. The key characteristic of a chronic event is the frequency factor. These chronic events happen over and over again for the same reason or mode. For instance, on a given pump failure, the bearing may fail three or four times a year, or if you a have a bottle-filling line and the bottles continuously jam, that would be considered a chronic event. Chronic events tend not to get the attention of sporadic events because their individual occurrences are usually not very costly. Therefore, rarely would we ever assign a dollar figure to an individual chronic event.

What most people fail to realize is the tremendous effect the frequency factor has on the cost of chronic failures. A stoppage on a bottling line due to a bottle jam may take only 5 minutes to correct when it occurs. If it happens five times a day, then we are looking at 152 hours of downtime per year. If an hour of downtime costs $10,000, then we are looking at a cost of approximately $1,520,000. As you can see, the frequency factor is very powerful. But since we tend to only see chronic events in their individual state, we sometimes overlook the accumulated cost. Just imagine if we were to go into a facility and aggregate all of the chronic events over a year's time and multiply their effects by the number of occurrences. The yearly losses would be staggering.

Let's take a look at how chronic and sporadic events relate to the discussion of problems and opportunities. Sporadic events by their definition take us below the status quo and tend to take an extended period of time to restore. When we restore we get back to the status quo. This is very much like what happens when we react to a problem. The problem occurs and we take some action to get back to the status quo. Chronic events, on the other hand, happen so routinely that they actually become part of the status quo or the job. Therefore, when they occur, they do not take us below our performance norm. If, in turn, we were to eliminate the chronic or repetitive events, then the elimination would actually cause the status quo to improve. This improvement is the equivalent of realizing an opportunity. So by focusing on chronic events and eliminating the causes, and not simply fixing the symptoms, we are really working on opportunities. As mentioned previously, when we work on opportunities we actually progress the organization (Figure 4.4).

Now that we know that eliminating chronic events can progress the organization, we have to look at the significance of chronic events. Sporadic events by their very nature are high-profile and high-cost events. But we can amortize those costs over a

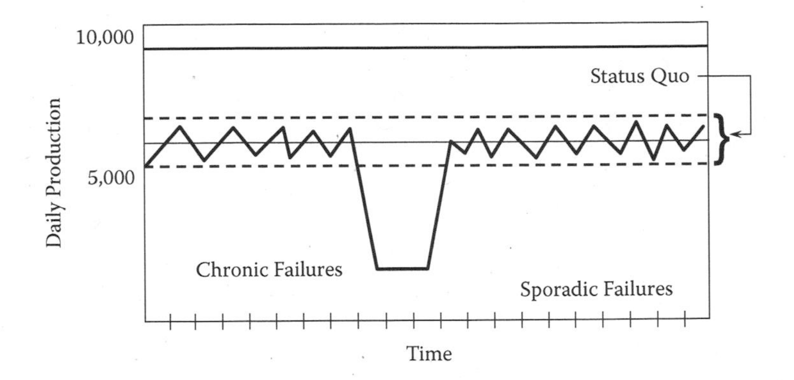

FIGURE 4.4 The linkage.

long period of time so the effect is not as severe. Consider if the engine in your car blew up and you had to replace it. To the average motorist this would be a sporadic event. But if we amortize the cost over the remaining life of the car, it becomes less of a burden. Chronic events, on the other hand, have a relatively low impact on an individual basis, but we often overlook their true impact. If we were to aggregate all of the chronic events from a particular facility and look at their total cost over a 1-year period, we would see that their impact is far more significant than any given sporadic event, simply due to the frequency factor.

Consider how all of the events actually affect the profitability of a given facility. As we all know, we are in business to make a profit. When a sporadic event occurs it affects the profitability of a facility significantly the year that it occurs, but once the problem has been resolved, profitability gets back to "normal." The dilemma with chronic events is that they usually don't ever get resolved, so they affect profitability year after year. If we were to eliminate such events instead of just reacting to their symptoms, we could make great strides in profitability. Imagine if we had 10 facilities and we were able to reduce the number of losses in order to obtain 10% more throughput from each of those facilities. In essence we would have the capacity of one new facility without spending the capital dollars. That is the power of resolving chronic issues.

Following is an example of a chronic event success story. In a large mining operation, the management wanted to uncover their most significant chronic events. In this operation, they have a large crane or "drag line" as they call it. This drag line mines the surface for the product. The product is then placed in large piles where a machine called a bucket wheel moves up and down the pile putting the product onto a conveyor system. This is where the product is taken downstream to another process of the operation. One day, one of the analysts was talking to one of the field maintenance representatives who said he spends a majority of time resetting conveyor systems whose safety trip cord is triggered. He estimated that this activity took anywhere from 10 to 15 minutes to resolve per trip. Now, this individual did not see this activity as a "failure" by any means. It was just part of the job he had to do. Upon further investigation, it was discovered that other people were also resetting tripped conveyors. By their estimation this was happening approximately 500 times a week to the tune of about $7,000,000 per year in lost production! Just by identifying this as an undesirable event allowed them to take instant corrective action. By adding a simple procedure of removing large rocks with a bulldozer prior to bucket wheel activity, approximately 60% of the problem went away. These types of stories are not uncommon. We get so ingrained in what we are doing that we sometimes miss the things that are so obvious to outsiders.

Similarly, in a hospital setting, we were looking at the number of times blood had to be redrawn in an emergency room of a 225-bed acute care hospital. At the conclusion of our Opportunity Analysis (to be discussed in detail in Chapter 5), we found that 10,013 blood redraws were taken in the last 12-month period. Next, we aggregated the average costs per blood redraw. These costs include things like the costs for syringes, gauze, tech time, transport time, opportunity costs for the real estate in the OR, etc. When compiled we found that, on average, each blood redraw was costing $300. The math is simple from this point on: we multiply 10,000 redraws times $300/redraw and we uncover a whopping $3,000,000 worth of hidden losses. On any individual occurrence no one sees this as a failure. It is viewed as a cost of doing business. This is the power of evaluating chronic failures.

To wrap this up we will end with yet another story. We were working with a major oil company that was trying to reduce its maintenance budget. So they hired our firm to teach them the methods being explained in this text. A manager opened the 3-day session by stating that he had been mandated by his superiors to significantly reduce the maintenance budget. He told them that the maintenance budget for this particular facility was approximately $250,000,000. He went on to explain that some analysis was done on the budget to find out how the money was being spent. It turns out that 85% of the money was spent in increments of $5,000 or less. So by his estimation he was spending about $212,000,000 in chronic maintenance losses. This was just maintenance cost, not lost production cost!

So he tells the 25 engineers in the training class that he has two options to reduce this maintenance cost (Table 4.1):

1. He can eliminate the need to do the work in the first place, or
2. He could just eliminate maintenance jobs.

He says that if they could eliminate the need to do the work in the first place (e.g., reduce the number of chronic or repetitive failures), then he felt they could reduce the maintenance expenditures by about 20%. This would be a savings of about $42,000,000. If they were really successful, they could eliminate 30% or $63,000,000.

He went on to say that "if I take Option 2 and let approximately 100 maintenance people go, that will probably net the company about $7,500,000 of which I will have the same, if not more, work and fewer people to have to address the additional work." To make a long story short, the people in the training class opted for Option 1, reducing the need to do the work using their abilities to solve problems!

So to sum up this discussion on failure classification, let's look at the key ideas presented. We live in a world of problems and opportunities. We would all love to take advantage of every opportunity that came about, but it seems as if there are

TABLE 4.1
Options to Reduce Maintenance Budget

	Scenerio: Oil Refinery Example		
Annual maintenance cost		$250,000,000	
Chronic losses		85.00%, increment of $5000 or less	
Total		$212,500,000	
Reduce the Need for the Work		**Option A**	
	20.00%	$42,500,000	**Net savings**
	30.00%	$63,750,000	**Net savings**
	40.00%	$85,000,000	**Net savings**
Reduce People		**Option B**	
Employees		1500	
Average loaded salary		$75,000	
Reduce employees by 7%:		105	
Net savings		**$7,875,000**	

too many problems confronting us to take advantage of the opportunities. A good way to take advantage in a business situation is to eliminate the chronic or repetitive events that confront us each and every day. By eliminating this expensive, non-value-added work, we are really achieving opportunities as well as adding additional time to eliminate more problems. In the next chapter, we will discuss a method for uncovering all of the events for a given process and delineating which of those events are the most significant from a business perspective.

RCA AS AN APPROACH

We mentioned this briefly in the Introduction, but it is appropriate to also mention it here. RCA is certainly applicable to both chronic and sporadic events in any industry. However, focusing on RCA as only an incident or accident tool does not optimize its potential for an organization. Using RCA in this fashion limits its effectiveness and treats it as an off-the-shelf tool for reactive situations.

When using RCA as an approach, we seek to break the paradigm that "chronic events" are an accepted cost of doing business because they are compensated for in the budget. We seek to solve these chronic events down to their root causes and pass the knowledge on to others in the organization who may be accepting them as a cost of doing business as well. This is the knowledge management and transfer component of RCA that we discussed earlier.

Also, many do not realize that the chronic types of events are actually precursors to the sporadic events. It is our experience that when reviewing the sporadic investigations in which we have been involved over the past 20 years, rarely do we find "revelations." Most of the time we find the true latent causes to be systems that are in place and have been the norm for some time. They have been chronically accepted over the years to the point that no one questions them anymore.

All it takes is one trigger, one decision, to make a chronic event a sporadic one! This was demonstrated on the space shuttle Challenger, as the O-ring design flaws were known from the beginning. That chronic problem existed for years and was an *acceptable risk* according to the flight readiness plan. In the Challenger Disaster Final Report this gradual deterioration of safety standards was referred to as *Normalization of Deviance*. Only when the decision was made to launch at 36°F (15°F colder than any other flight) did that chronic failure become a sporadic one. Bridging this to our working environments, can't this happen to us? Doesn't this happen to us?

5 Opportunity Analysis
"Mindfulness"

Mindfulness is about the ability of a system to concentrate on what is going on here and now.*

With all the noise and distraction of a reactive work environment it is sometimes easy to overlook the obvious. For instance, if we wanted to perform a Root Cause Analysis (RCA) on an event, would we know which event was the most significant or costly? Experience demonstrates that we would not. In a reactive environment, we naturally become focused on the short term. We tend to look at the problems or events that just happened and naturally think they are the most significant at the time. This is a problem because what happened yesterday, in most cases, is not the most significant or compelling issue today. We need to take a more macro look at the situation. For these reasons we must depend on the strategy development process described earlier to ensure that we are working on the events that truly add value to the "bottom line" of the business.

In order to determine where our most significant issues are we should employ techniques that will allow us to objectively look at all the historical events contributing to our performance or lack thereof. Failure Modes and Effects Analysis, or FMEA, was developed in the aerospace industry to determine what failure events *could* occur within a given system (e.g., a new aircraft) and what the associated effects would be if those events did indeed occur. This technique, albeit effective, is very man-hour intensive. It is estimated that a typical FMEA in the aerospace industry takes numerous man-years to perform. There are many good reasons why this technique takes so long to perform as well as significant benefits to this industry. However, this technique is far too laborious to be performed in most industries, such as the process and discrete manufacturing sectors. Therefore, we had to take the basic concept and make it more "industry friendly." When discussing this *modified* FMEA technique, we will refer to it as Opportunity Analysis or OA.

Before we continue with the discussion on how to develop an Opportunity Analysis, let's first talk about why you would want to perform one in the first place. There really are two basic reasons to perform an Opportunity Analysis. The first and foremost is to make a legitimate business case to analyze one event versus another. In other words, an OA creates the financial or business reason to show a listing of all the events within a given organization or system and delineate in dollars and cents why you are choosing one issue versus another. It allows the analyst to speak in the "language of business."

* Weick, Karl E. and Kathleen M. Sutcliffe. 2007. *Managing the Unexpected: Resilient Performance in an Age of Uncertainty*. San Francisco: Jossey-Bass, p. 35.

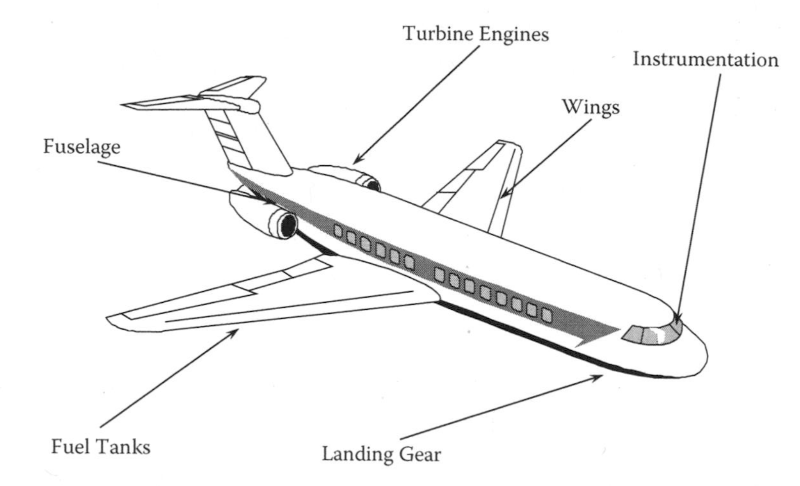

FIGURE 5.1 Aircraft subsystem diagram.

The second compelling reason is to focus the organization on what the most significant events really are so that quantum leaps in productivity can be made with fewer of the organization's resources being utilized. Experience again has shown that the Pareto Principle* works with such events just like it does in other areas. It goes something like this: 20% or less of the undesirable events that we uncover by conducting an in-depth Opportunity Analysis will represent approximately 80% of the losses for that organization. You may have heard this also called the 80/20 rule. We will discuss the 80/20 rule later in this chapter.

As mentioned previously, the FMEA technique was developed in the aerospace industry, and we will refer to this as the "traditional" FMEA method. Modifications are necessary to make the traditional FMEA more applicable in other organizations. Therefore, based on the modifications that we will explain in this chapter, we will call this technique Opportunity Analysis. The key difference between the two methods is that the traditional method is probabilistic, meaning it looks at what could happen. In contrast, Opportunity Analysis looks only at historical events. We only list items that have actually happened in the past. For the historical method, we are not exactly interested in what might happen "tomorrow" as we are in what did happen "yesterday."

Let's take a look at a simple example of both a traditional FMEA and an Opportunity Analysis. Our intention is not so much to develop experts in traditional FMEA as it is to give a general understanding of how FMEA and thus Opportunity Analysis was derived. In the aerospace industry, we would perform a traditional FMEA on a new aircraft that is being developed. So the first thing we would do is to break the aircraft down into smaller subsystems. A typical aircraft would have many subsystems such as the wing assembly, instrumentation system, fuselage, engines, etc. (Figure 5.1).

* PROACT RCA Methods Course is copyrighted by Reliability Center, Inc., Hopewell, VA.

TABLE 5.1
Traditional FMEA Sample

Subsystem	Mode	Effects on Other Items	Effects on Entire System	Severity	Probability	Criticality
Turbine Engine	Cracked Blade	If blade releases, it could fracture other blades	Loss of one engine, reduced power and control	8	0.02	0.16

From there the analysis would look at each of the subsystems and determine what failure modes might occur and if they did, "What would be their effects?" Let's take a look at a simple example in Table 5.1.

In Table 5.1, we begin by looking at the Turbine Engine subsystem. We begin listing all of the potential failure modes that might occur on the turbine engines. In this case, we might determine that a turbine blade could fracture. We then ask what the effects on other items within the turbine engine subsystem might be. If the blade were to release, it could fracture the other turbine blades. The effects on the entire system, or the aircraft as a whole, would be loss of the engine and reduced power and control of the aircraft. We then begin examining the severity of the failure mode. We will use a simple scale of 1 to 10 where 1 is the least severe and 10 is the most severe. We have simplified this for explanation purposes, but a traditional FMEA analyst would have specific criteria for what constitutes gradients of severity. In this example, we will say that losing a turbine blade would constitute a severity of "8." Now comes the probability rating. We would have to collect enough data to determine the relative probability of this occurrence based on the design of our aircraft. We will assume that the probability in this case is 0.02 or 2%. The last step is to multiply the severity times the probability to get a criticality rating. In this case the rating would be calculated as follows:

$$\begin{gathered} 8 \times 0.02 = 0.016 \\ \text{Severity} \times \text{Probability} = \text{Criticality} \end{gathered} \tag{5.1}$$

EQUATION 5.1 Sample criticality equation.

This means that this line item in the FMEA has a criticality rating of 0.016. We would then repeat this process for all of the failure modes in the turbine engines and all of the other major subsystems.

Once all of the items have been identified, it is time to prioritize. We would sort our criticality column in descending order so that the largest criticality ratings would bubble up to the top and the smaller ones would fall to the bottom. At some point the analyst would make a cut specifying that all criticalities below a certain number are delineated as an acceptable risk, and all above need to be evaluated to determine a way to reduce the severity and, more importantly, the probability of occurrence.

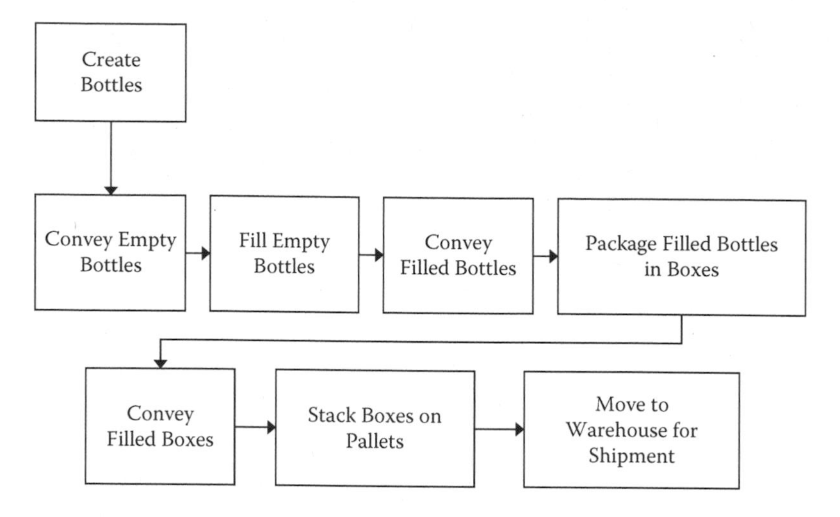

FIGURE 5.2 Sample lubricants plant.

Bear in mind that this is a long-term process. A great deal of attention is placed on determining all of the possible failure modes, and even greater attention is paid to substantiating the severity and probability. Thousands of hours are spent running components to failure to determine probability and severity. Computers, however, have helped in this endeavor, in that we can simulate many occurrences by building a computer model and then playing "what if" scenarios to see what the effects would be.

We do not have the time or resources in business, healthcare, and industry to perform a thorough traditional FMEA on every system. Nor does it make economic sense to do so on every system. What we have to do is modify the traditional FMEA process to help us uncover the problems and failures that are currently occurring. This allows us the ability to see what the real cost of these problems are and how they really affect our operation. Let's look at a simple example.

Consider that we are running a lubricants plant. In this plant we are doing the following (Figure 5.2):

1. Providing the plastic bottles for the lubricant
2. Conveying the bottles to the filling machine to be filled with lubricant
3. Conveying the filled bottles to the packaging process to be boxed in cases
4. Conveying the filled boxes to be put onto pallets
5. Moving the pallets to the warehouse where they await shipping

The next step is to determine all of the undesirable events that are occurring in each of our subsystems. For instance, if we were looking at the Fill Empty Bottles subsystem, we would uncover all of the undesirable events related to this subsystem. Let's look at a simple example (Table 5.2).

The idea is to delineate the events that occurred that caused an upset in the Fill Empty Bottles subsystem. In this case, one of the events would be a bottle stoppage. The mode of this particular event is that a bottle became jammed in the filling

TABLE 5.2
Opportunity Analysis Line Item Sample

Subsystem	Event	Mode	Frequency/Yr.	Impact	Total Loss
Fill Empty Bottles	Bottle Shortage	Bottle Jam	1000	$150	$150,000

cycle. It occurs approximately 1000 times a year or about three times a day. The approximate impact for each occurrence is $150 in lost production. If we multiply the frequency times the impact for each occurrence, we would come to a total loss of $150,000 per year.

If we were to continue the analysis, we would pursue each of the subsystems, delineating all of the events and modes that have caused an upset in their respective subsystems. The end result would be a listing of all the items that contribute to lost production and their respective losses. Based on that listing, we would select the events that were the greatest contributors to lost production and perform a disciplined Root Cause Analysis (RCA) to determine the root causes for their existence.

Now that we understand the overall concept of FMEA, let's take a detailed look at the steps involved in conducting an Opportunity Analysis. There are seven basic steps:

1. Perform preparatory work
2. Collect the data
3. Summarize and encode results
4. Calculate loss
5. Determine the "Significant Few"
6. Validate results
7. Issue a report

STEP 1—PERFORM PREPARATORY WORK

As with any analysis, there is a certain amount of preparation work that has to take place. Opportunity Analysis is no different in that it also requires several up-front tasks. In order to adequately prepare to perform an Opportunity Analysis you must accomplish the following tasks:

- Define the system to analyze.
- Define the undesirable event.
- Draw a block diagram (use contact principle).
- Describe the function of each block.
- Calculate the "gap."
- Develop preliminary interview sheets and schedule.

DEFINE THE SYSTEM TO ANALYZE

Before we can begin generating a list of problems, we have to decide which system to analyze. This may sound like a simple task, but it does require a fair amount of thought on the analyst's part. When we teach this method to our students, their usual response is to take an entire facility and make it the system. This is a prescription for disaster. Trying to delineate all of the failures and/or problems in a huge oil refinery, for instance, would be a daunting task. What we need to do is localize the system down to one system within a larger system. For instance, a large oil refinery is comprised of many operating units. There is a Crude Unit, Fluid Catalytic Cracking Unit (FCCU), Delayed Coking Unit (DCU), and many others. The prudent thing to do would be to select one unit at a time and make that unit the focus of the analysis. For example, the Crude Unit would be the system to study and then we would break the Crude Unit in to many subsystems. In other words, we should not bite off more than we can chew when selecting a system to study. We have seen many cases where analysts first do a rough cut to see which area of the facility either comprises a bottleneck or is expending the greatest amount of expense.

Define Undesirable Event

This may sound a little silly, but we have to define exactly what an "undesirable event" is in our facility. During every seminar that we teach on this subject, we ask students to write down their definition of an undesirable event at their facility. Just about every time, every student has a different definition. The fact is, if we are going to collect event data, everyone involved must be using a consistent definition. If we are collecting event data and there is no standardized definition, then everyone will give us their perceptions of what undesirable events are occurring in their work areas. For instance, if we ask a machine operator what undesirable events he sees, he will probably give us processing-type events, a maintenance mechanic will probably give us machinery-related events, whereas a safety engineer would probably give all of the safety issues. The dilemma here is that we lose focus when we do not have a common definition of an undesirable event.

The key to making an effective definition of an undesirable event is to ensure that the definition coincides with a particular business objective specified in the strategy map. For example, if we are in a sold-out position and our objective is to increase production utilization, then our definition should be based primarily around continuous production or limiting downtime. Let's take a look at some common definitions that we have run across over the years. Some are pretty good and others are unacceptable. An undesirable event is

- Any loss that interrupts the continuity of maximum quality production
- A loss of asset availability
- The unavailability of equipment
- A deviation from the status quo
- One that does not meet target expectations
- Any secondary defect

The first definition of an undesirable event, "any loss that interrupts the continuity of maximum quality production," is a pretty good definition and one that we see and use quite frequently. Let's analyze this definition. In most manufacturing facilities, we often take our processes off line to do routine maintenance. The question when we take these planned shutdowns becomes, "Are we experiencing an undesirable event based on the first definition above?" The answer is an emphatic YES! The definition states that any loss that interrupts the continuity of maximum quality production is deemed an undesirable event. Even if we plan to take the machines out of service, it still interrupts the continuity of maximum quality production. Now, we are not saying that we should not make periodic shutdowns for maintenance reasons. All we are suggesting is that we look at them as undesirable events so that we can analyze if there is any way to stretch out the intervals between each planned shutdown and reducing the amount of time a planned shutdown actually takes. For instance, in many industries, we still have what we call "Annual Shutdowns." How often do we have an "Annual Shutdown"? Every year, of course! It says so right in the name. Obviously, the government and other legislative bodies regulate some shutdowns such as pressure vessel inspections. But in many cases, we are doing these yearly shutdowns just because the calendar dictates it. Instead of performing these planned shutdowns on a time basis, maybe we should consider using a condition basis. In other words, let the condition of the equipment dictate when a shutdown has to takes place.

This idea of looking at planned shutdowns as an undesirable event is not always obvious or popular. But if we are in a sold-out position, we must look at anything that takes us away from our ability to run 8760 hours a year at 100% throughput rate. Now let's consider a different scenario. In many facilities, we have spare equipment, just in case the primary piece of equipment fails. It is sort of an insurance policy for unreliability. In this scenario, if the primary equipment failed and the spare equipment "kicked in," would this interrupt the continuity of maximum quality production? Providing the spare functions properly, the answer here would have to be NO. Since we had the spare equipment in place and operating, we did not lose the production. That event would not end up on our list because it did not meet our definition of an undesirable event. This is also a hard pill for some of us to swallow. But that is the tough part about focusing. Once we define what an undesirable event is, we must list only the events that meet that definition.

Let's consider the definition, "a deviation from the status quo." This definition has many problems. The primary problem is, "What happens if you have a positive deviation?" Should that be considered a failure? Probably not. How about the words "status quo"? For one thing, status quo is far too vague. If we were to ask 100 people to describe the status quo of the United States today, they would all give us a different answer. In addition, the status quo does not always mean that things are good—it just says that things are the way they are. If we were to rewrite that definition, it would make more sense if it looked like this:

> An undesirable event is a negative deviation from 1 million units per day.

So why bother with a definition? It serves multiple purposes. First of all, we cannot perform an Opportunity Analysis without it. But in our opinion, that is the least important reason. The biggest advantage of an agreed-upon definition is that it fosters precise communication between everyone in the facility. It gets people focused on the most important issues. In short, it focuses people on what is really important and the fact that we are adhering to the strategy defined in the strategy map.

When we devise a definition of an undesirable event, we need to make sure that it is short and to the point. We certainly would not recommend a definition that is several paragraphs long. A good definition can and should be about one sentence. Our definition should only address one business objective at a time. For example, a definition that states, "An undesirable event is anything that causes downtime, an injury, an environmental excursion, and/or a quality defect" is trying to capture too many objectives at one time, which in turn will cause the analysis to lose focus. If we feel the need to look at each of those issues, then we need to perform separate analyses for each of them. It may take a little longer, but we will maintain the integrity of the analysis.

Last but not least, it is important to get decision makers involved in the process. We would recommend having someone in authority sign off on the definition to give it some credence and clout. If we are lucky, the person in authority will even modify the definition. This will, in essence, create buy-in from that person.

Drawing a Process Flow Diagram or Block Diagram (Use the Contact Principle)

Now that we have defined the system to analyze and the definition of an undesirable event that is most appropriate, we now have to create a simple flow diagram of the system being analyzed. This diagram will serve as a job aid later when we begin collecting data. The idea of a diagram is to show the flow of product from point A to point B. We want to list out all of the systems that come in contact with the product. Let's refer to our lubrication facility example (Figure 5.3).

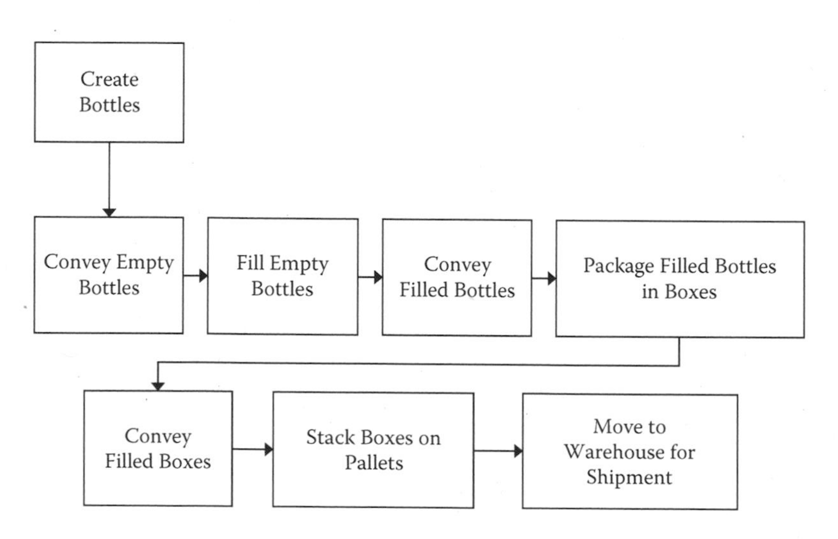

FIGURE 5.3 Block diagram example.

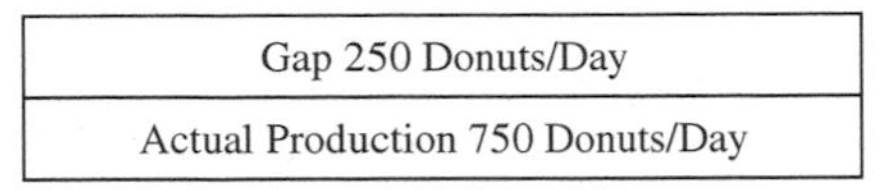

FIGURE 5.4 Sample gap analysis.

Each of these blocks indicates a subsystem that comes in contact with the product. We use this diagram to help us graphically represent a process flow to which it is easy to refer. Many facilities maintain such detailed drawings and use them on a daily basis. Oftentimes such diagrams are referred to as Process Flow Diagrams, or PFDs. If we have such diagrams already in our facilities, we are ahead of the game. If we do not, we must simply create a simple diagram like the one in Figure 5.3 to help represent the overall process. We will discuss how to use both the undesirable event definition and the contact flow diagram in the data collection phase.

Describe the Function of Each Block

In some cases, drawing the block diagram in itself is not enough of an explanation. We may possibly be working with some individuals who are not intimately aware of the function of each of the systems. In these cases, it will be necessary for us to add some level of explanation for each of the blocks. This will allow those who are less knowledgeable in the process to participate with some degree of background in the process.

Calculate the "Gap"

In order to determine success, it will be necessary to demonstrate where we are as opposed to where we could be. In order to do this, we will need to create a simple gap analysis. The gap analysis will visually show where we currently are versus where we could be. For instance, let's assume that we have a donut machine that has the potential of making 1000 donuts per day, but we are only able to make 750 donuts per day. The gap is 250 donuts per day. We will use our Opportunity Analysis to uncover all of the reasons that are keeping us from reaching our potential of 1000 donuts per day (Figure 5.4).

Develop Preliminary Interview Sheets and Schedule

The last step in the preparatory stage is to design an interview sheet that is adequate to collect the data consistent with your undesirable event definition and to set up a schedule of people to interview to get the required data. Let's look at the required data elements or fields. In every analysis we will have the following data elements (Table 5.3):

- Subsystem—This correlates to the blocks in our block diagram.
- Event—The event is the actual undesirable event that matches the definition we created earlier.

TABLE 5.3
Sample Opportunity Analysis Worksheet

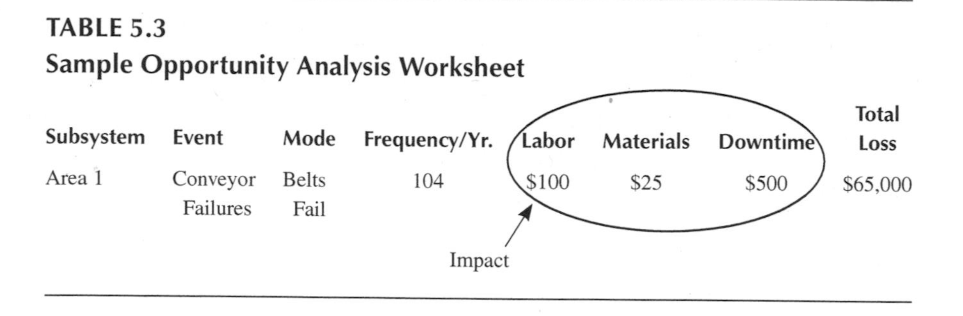

Subsystem	Event	Mode	Frequency/Yr.	Labor	Materials	Downtime	Total Loss
Area 1	Conveyor Failures	Belts Fail	104	$100	$25	$500	$65,000

- Mode—The mode is the apparent reason that the undesirable event exists.
- Frequency per Year—This number corresponds to the number of times the mode actually occurs in a year's time.
- Impact per Occurrence—This figure represents the actual cost of the mode when it occurs. For instance, we will look at materials, labor, lost production, fines, scrap, etc. This data element can represent any item that has a determinable cost.
- Total Loss per Year—This is the total loss per year for each mode. It is calculated by simply multiplying the frequency per year by the impact per occurrence.

In order to develop an effective interview sheet we have to create it based on our definition. The first four columns (subsystem, event, mode, and frequency) are always the same. The impact column, however, can be expanded on to include whatever cost elements we feel are appropriate for the given situation. For instance, some do not include straight labor costs since we have to pay such costs regardless. We will, however, include any overtime costs associated with the mode since we would not have incurred the expense without the event occurring.

The last item in the preparatory stage is to determine which individuals we should interview and to create a preliminary interview sheet to list all of the individuals to talk to in order to collect this event information. We will further discuss what types of people to interview in the next topic.

STEP 2—COLLECT THE DATA

There are a couple schools of thought when it comes to how to collect the data that is necessary to perform an Opportunity Analysis. On one side, there are those who believe that all data can be retrieved from a computerized system within the organization. The other side believes that it would be virtually impossible to get the required data from an internal computer system since the data going into the system is suspect at best. Both sides are correct to some degree. Our data systems do not always give the precise information that we need, although they can be useful to verify trends that would be uncovered by interviewing people.

We will explore both of these alternatives in this chapter and the next. However, the analyst leading the Opportunity Analysis will ultimately be responsible for making the decision as to whether the more accurate and timely data comes from the people or the existing information system. In this chapter we will continue on with the manual approach of collecting data from the raw source—the people. In the next chapter, we will explore the data collection opportunities that are available from an Asset Performance Management (APM) system, thus automating the effort.

It is recommended that when using the manual method of data collection (interviewing technique) we take a two-track approach. We begin collecting data from people through the use of interviews. We use the interviews very loosely, as we will explain later. Once we have collected and summarized the interview data, we can use our existing data systems to verify financial numbers and see if the computer data supports the trends that we uncovered in our interviews. The numbers will not be the same but the trends may very well be. So let's discuss how to go about collecting event data using an interview method.

As mentioned in our previous discussion, we developed two job aids. We had an undesirable event definition and a block diagram of the process flow. We are now going to use those two documents to help us structure an interview. We begin the interview by asking the interviewee to delineate any events that meet our definition of an undesirable event within a certain subsystem. This creates a focused interview session. As we said earlier, an interview generally has a kind of negative connotation. In order to gain employment we typically have to go through an interview, which is sometimes a stressful situation. We often watch TV police shows where a suspect is being interviewed (i.e., interrogated) in a dark, smoky room. We would choose to make our interviews much more informal. Think of them more as an information-gathering session instead of a formal interview. This will certainly improve the flow of information.

Now, who would be good candidates to talk to in an interview or discussion session? It is important to make sure that we have a good cross-section of people to talk to. For instance, we would not want to talk to just maintenance personnel because we may only get maintenance-related information. So what we should strive to do is interview across disciplines, meaning that we get information from maintenance, operations, technical, and even administrative personnel. Only then will we have the overall depth that we are seeking. There is also the question of what level of person we want to talk with. In most organizations there is a hierarchy of authority and responsibility. For instance, in a manufacturing plant there are the hourly or field-level employees who are primarily responsible for operating and maintaining the day-to-day operations to keep the products flowing. Then there is a middle supervisory level that typically supervises the craft and operator levels. Above the supervisory levels are the management levels that typically look at the operation from a more global perspective.

When trying to uncover undesirable events and modes it makes sense to go to the source. This means talking to the people closest to the work. In most cases, this would describe the hourly workforce. They deal with undesirable events each and every day and are usually the ones responsible for fixing those problems. For this reason, we would recommend spending most of the interview time with people at this level. If we think about it, the hourly workforce is the most abundant resource that rarely, in our experience, is used to its fullest potential. Sometimes getting this wealth of knowledge is as easy as just asking for it. We are certainly not suggesting that we should not talk with supervisory level employees or above. They also have a vast amount of experience and knowledge of the operation. As far as upper-level managers go, they usually have a more strategic focus on the operation. They may not have the specific information required to accomplish this type of analysis. There are exceptions to every rule, however. We once worked at a facility where the Plant Manager routinely would log into the Distributive Control System (DCS) from his home computer in the middle of the night to observe the actions of his operators. When they made an adjustment that he thought was suspect, he would literally call the operator in the control room to ask why they did what they did. Imagine trying to operate in such a micromanaged environment? Although we do not support this manager's practice, he probably would have a great deal to offer in our analysis of process upsets because he had intricate knowledge of the process itself.

Another idea that we have found to be very useful when collecting event information is to talk to multiple people at the same time. This has several benefits. For one, when a person is talking it is spurring something in someone else's mind. It also has a psychological effect. When we ask people about event information, it may be perceived as a "witch hunt." In other words, they might feel like management is trying to blame people. By having multiple interviewees in a session, it appears to be more of a brainstorming session instead of an interrogation.

The interviewing process, as we have learned over the years, is really an art form more than a science. When we first started to interview, we soon learned that it can sometimes be a difficult task. It is like golf—the more we practice proper technique the better the final results will be. An interview is nothing more than getting information from one individual to another as clearly and accurately as possible. To that end, here are some suggestions that will help you to become a more effective interviewer. Some of these are very specific to the Opportunity Analysis process, but others are generic in that they can be applied to any interviewing session.

- Be very careful to ask the exact same lead questions to each of the interviewees. This will eliminate the possibility of having different answers depending on the interpretation of the question. Later we can expand on the questions, if further clarification is necessary. We can use our undesirable event definition and block flow diagram to keep the interviewees focused on the analysis.
- Make sure that the participants know what an Opportunity Analysis is, as well as the purpose and structure of the interviews. If we are not careful, the process may begin to look more like an interrogation than an interview to the interviewees. An excellent way to make our interviewees comfortable with the process is to conduct the interviews in their work environments instead of ours. For instance, go to the break area or the shop to talk to these people. People will be more forthcoming if they are comfortable in their surroundings.
- Allow the interviewees to see your notes. This will set them at ease since they can see that the information they are providing is being recorded

accurately. NEVER use a tape recorder in an Opportunity Analysis session because it tends to make people uncomfortable and less likely to share information. Remember, this is an information-gathering session and not an interrogation.

- If we do not understand what someone is telling us, let them use a pen to draw a simple diagram of the event for further understanding. If we still do not understand what they are trying to describe, then we should go out to the actual work area where the problem is occurring so that we can actually visualize the problem.
- Never argue with an interviewee. Even if we do not agree with the person, it is best to accept what he or she is saying at face value and double check it with the information from other interviews. The minute we become argumentative, it reduces the amount of information that we can get from that person. Not only will that person not give us any more information, chances are he or she will alert others to the argument and they will not want to participate either.
- Always be aware of interviewees' names. There is nothing sweeter to a person's ears than the sound of his or her own name. If you have trouble remembering, simply write the names down in front of you so that you can always refer to them. This gives any interview or discussion a more personal feel.
- It is important to develop a strategy to draw out quiet participants. There are many quiet people in our workforce who have a wealth of data to share but are not comfortable communicating it to others. We have to make sure that we draw out these quiet interviewees in a moderate and inquiring manner. We can use nominal group techniques where we ask each of the people with whom we are talking to write their comments down on an index card and then compile the list on a flip chart. This gives everyone the same chance to have their comments heard.
- Be aware of body language in interviewees. There is an entire science behind body language. It is *not* important that we become an expert in this area. However, it is important to know that a substantial portion of human communication is made through body language. Let the body language talk to us. For instance, if someone sits back in a chair with his or her arms firmly crossed, the person may be apprehensive and not feel comfortable providing the information that we are asking for. This should be a clue to alter our questioning technique to make that person more comfortable with the situation.
- In any set of interviews, there will be a number of people who are able to contribute more to the process than others. It is important to make a note of the extraordinary contributors so that they can assist us later in the analysis. They will be extremely helpful if we need additional event information for validating our finished Opportunity Analysis, as well as assisting us when we begin our actual Root Cause Analysis (RCA).
- Remember to use the undesirable event definition and block diagram to keep interviewees on track if they begin to wander off of the subject.

- We should strive to keep interview sessions relatively short. Usually about 1 hour is suitable for an interview session. This process can be very intensive and people can become tired and sometimes lose their focus. This is dangerous because it begins to upset the validity of the data. So as a rule, 1 hour of interviewing is plenty.

STEP 3—SUMMARIZE AND ENCODE DATA

At this stage, we have generated a vast amount of data from our interviews. We now have to begin summarizing this information for accuracy. While conducting our interviews, we will be getting some redundant data from different interviewees. For instance, a person from the night shift might be giving us the same events that the day shift person gave us. So we have to be very careful to summarize the information and encode it properly so that we do not have redundant events and are essentially "double dipping."

The easiest way to collect and summarize the data is to input it into an electronic spreadsheet or database like Microsoft® Excel* or Microsoft® Access.† Of course, we could certainly do this manually with a pencil and paper, but if we have a computer available, we should take the opportunity to use it. It will save many hours of frustration with performing the analysis manually. Once we have input all of the information into our spreadsheet, we now have to look for any redundancy. We should always remember to use a logical coding system when inputting information into a computer. Once we define what that logical coding system is, stick with it. Otherwise, the computer will be unable to provide the results we are trying to achieve. Let's take a look at the example in Table 5.4 to help us understand logical coding.

If we were to use the coding portrayed on the bottom of the graphic we would get inconsistent results when we tried to summarize the data. Therefore, we have to

TABLE 5.4
Logical and Illogical Coding

Subsystem	Failure Event	Failure Mode
	Illogical Coding	
Area 6	Pump 102 Failure	Bearing Fails
Area 6	Pump 102 Failure	Seal Fails
Area 6	Pump 102 Failure	Motor Fails
	Logical Coding	
In Area 6	Pump 102 Failure	Bearing Break
Area 6	Failure of CP—102	Seals
Area 6	Pump Failure—102	Failure of Motor

* Microsoft Excel is a registered trademark of the Microsoft Corporation.
† Microsoft Access is a registered trademark of the Microsoft Corporation.

TABLE 5.5
Example of Summarizing and Encoding

Subsystem	Event	Mode	Frequency	Impact
Recovery	Recirculation Pump Fails	Bearing Locks Up	12	12 hours
Recovery	Recirculation Pump Fails	Oil Contamination	6	1 day
Recovery	Recirculation Pump Fails	Bearing Fails	12	12 hours
Recovery	Recirculation Pump Fails	Shaft Fracture	1	5 days

TABLE 5.6
Example of Summarizing and Encoding Results

Subsystem	Event	Mode	Frequency	Impact
Recovery	Recirculation Pump Fails	Bearing Problems	12	12 hours
Recovery	Recirculation Pump Fails	Shaft Fracture	1	5 days

strive to use a coding system like the one depicted in the top of the graphic, which should give the required result when summarizing the data.

Now, how can we eliminate the redundant information that is given in the interview sessions? The easiest way is to take the raw data from our interviews and input it into our spreadsheet program. From there we can use the powerful sorting capability of the program to help look for the redundant events. The first step is to sort the entire list by the subsystem column. Then within each subsystem, we will need to sort the failure event column. This will group all of the events from a particular area so that we can easily look for duplicate events. Once again, if we do not use logical coding this will not be effective. So we should strive to be disciplined in our data entry efforts.

Let's take a look at Table 5.5 for an example of how to summarize and encode events. In this example, we are looking at the Recovery Subsystem and we have sorted by the Recirculation Pump Fails. Four different people at four separate times described these events. Is there any redundancy? The easiest way to see is to look at the modes. In this case we have two that mention the word bearing. The second is oil contamination. The interviewee was probably trying to help us out by giving us his opinion of the cause of the bearing failures. So in essence the first three events are really the same event. Therefore we will have to summarize the three events into one. Table 5.6 shows what it might look like after we summarize the items.

STEP 4—CALCULATE LOSS

Calculating the individual modes is a relatively simply process. The idea here is to multiply the frequency per year times the impact per occurrence. So if we have a mode that costs $5,000 per occurrence and it happens once a month, we have a $60,000-a-year problem. We usually choose to use financial measurements

TABLE 5.7
Example of Calculating the Loss

Event	Mode	Frequency	Impact	Total Lost Units	Total Loss
Pump Failure	Bearing Problems	12	500	6,000	$180,000
Off Spec. Product	Wrong Color	52	400	20,800	$624,000
Conveyor Failures	Roller Failures	500	50	25,000	$750,000

(e.g., dollars, Euros) to accurately determine loss. We may find that using another metric is a more accurate measurement for our business. For instance, we may want to track pounds, tons, number of defects, etc. But if it is possible, we should try to convert our measurement into financial currency. Money is the language of business and is usually the easiest way to communicate to all levels of the organization.

Let's look at a few examples of calculating the loss in Table 5.7. In this example, we are simply multiplying the frequency per year times the impact per occurrence, which in this case is in number of units. In other words, when each of these modes occurs, the impact is the number of units lost as a result. Notice that the last column is total loss in dollars. We simply multiply the number of lost units by the cost of each unit to give a total loss in dollars. That's all there is to it!

STEP 5—DETERMINE THE "SIGNIFICANT FEW"

We now have to determine which events out of all the ones we have listed are significant. We have all heard of the 80/20 rule, but what does it really mean? This rule is sometimes referred to as the Pareto Principle. The name Pareto comes from the early twentieth-century Italian economist who once said, "In any set or collection of objects, ideas, people, and events, a FEW within the sets or collections are MORE SIGNIFICANT than the remaining majority." This rule or principle demonstrates that in our world, some things are more important than others. Let's look at a few examples of this rule in action:

- Banking industry—In a bank approximately 20% or less of the customers account for approximately 80% or more of the assets in that bank.
- Hospital industry—In a hospital approximately 20% or less of the patients get 80% or more of the care in that hospital.
- Airline industry—20% or less of the passengers account for 80% or more of the revenues for the airline.

It also works in industrial applications. Throughout our years of experience and our clients, the rule holds true. Twenty percent or less of the identified events typically represents 80% or more of the resulting losses. This is truly significant if you think about it. It says that if we FOCUS on and eliminate the 20% of the events that

TABLE 5.8
Sample Opportunity Analysis Worksheet

Subsystem	Event	Mode	Frequency	Impact ❶	Total Loss	
Subsystem A	Event 1	Mode 11	30	$40,000	$1,200,000	
Subsystem A	Event 2	Mode 7	4	$230,000	$920,000	❺
Subsystem B	Event 3	Mode 1	365	$1,350	$492,750	
Subsystem A	Event 2	Mode 5	10	$20,000	$200,000	
Subsystem A	Event 2	Mode 8	10	$10,000	$100,000	
Subsystem B	Event 5	Mode 6	35	$2,500	$87,500	❹
Subsystem B	Event 4	Mode 4	1000	$70	$70,000	
Subsystem A	Event 4	Mode 12	8	$8,000	$64,000	
Subsystem B	Event 6	Mode 10	6	$8,000	$48,000	
Subsystem C	Event 4	Mode 13	4	$7,500	$30,000	
Subsystem B	Event 4	Mode 9	10	$2,500	$25,000	
Subsystem A	Event 1	Mode 2	12	$2,000	$24,000	
Subsystem A	Event 1	Mode 3	9	$2,500	$22,500	
Subsystem C	Event 6	Mode 14	6	$3,500	$21,000	
Total Loss					$3,304,750	❷
Significant Few Losses (Total Loss × .80)					$2,643,800	❸

represent 80% of the losses, we will achieve tremendous improvement in a relatively short period of time. It just makes common sense!

Think about how the rule applies to everyday living. We are all probably guilty of wearing 20% or less of the clothes in our closet 80% of the time. We all probably have a toolbox in which we use 20% of the tools 80% of the time. We spend all that money on all those exotic tools and most repairs require the screwdriver, hammer, and a wrench! We are all guilty of this! The rule even applies to business. Take, for instance, a major airline as described previously. It is not the once-a-year vacationer who generates most of the airline's revenue. It is the guy who flies every Monday morning and returns every Friday afternoon. So it makes sense that very few of the airline's customers represent most of its revenue and profits. Have you ever wondered why Frequent Flyer programs are so important to an airline? They know to whom they have to cater.

Let's take a look at the example in Table 5.8 to determine exactly how to take a list of events and narrow it down to the "Significant Few."

Step 1—Multiply the frequency column times the impact column to get a total annual loss figure.

Step 2—Sum the total annual loss column to obtain a global total loss figure for all the events in the analysis.

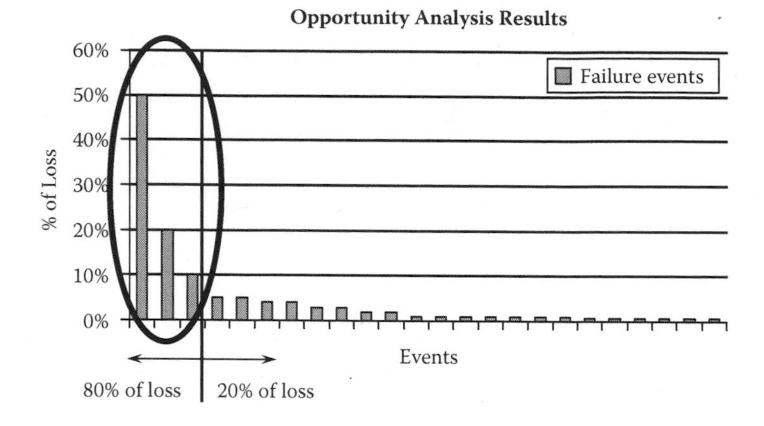

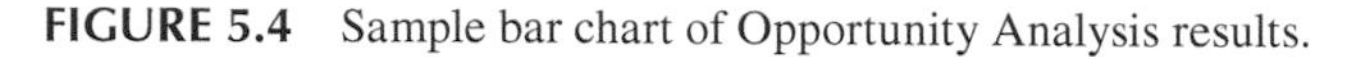

FIGURE 5.4 Sample bar chart of Opportunity Analysis results.

Step 3—Multiply the global total loss figure from Step 2 by 80% or 0.80. This will give us the "Significant Few" losses amount.
Step 4—Sort the total loss column in descending order so that the largest events bubble up to the top.
Step 5—Sum the total loss amounts from biggest to smallest until you reach the "Significant Few" loss amount.

In order to get the maximum effect, it is always wise to present this information in alternate forms. The use of graphs and charts will help you to effectively communicate this information to others around you. Figure 5.4 is a sample bar chart that takes the spreadsheet data and converts it into a more understandable format.

STEP 6—VALIDATE RESULTS

Although our analysis is almost finished, there is still more to accomplish. We have to verify that our findings are accurate. Our Opportunity Analysis total should be relatively close to our gap that we defined in our preparatory phase. The general rule is plus or minus 10% of the gap.

If we are way under that gap, we have either missed some events, undervalued them, or we do not have an accurate gap (actual versus potential). If we were to overshoot the gap, we probably did not do as good a job at removing the redundancies or we have simply overvalued the loss contribution.

At a minimum, we must double-check our "Significant Few" events to make sure we are relatively close. We do not look for perfection in this analysis simply because it would take too long to accomplish, but we do want to be close. This would be a good opportunity to go to our data sources like our Computerized Maintenance Management System (CMMS) or our Distributive Control System (DCS) to verify trends and financial numbers. Incidentally, if there were ever a controversy over a financial number it would be prudent to use numbers that the accounting department deems accurate. Also, it is better to be conservative with our financials so we do not risk losing credibility for an exaggerated number. The numbers will be high enough on their own without any exaggeration. Other verification methods might be more interviews or designed experiments in the field to validate interview findings. All in all, we want to be comfortable enough to present these numbers to anyone in the organization and feel that we have enough supporting information to back them up.

STEP 7—ISSUE A REPORT

Last, but certainly not least, we have to communicate our findings to decision makers so that we can proceed with solving some of these pressing issues. Many of us falter here because we do not take the time to adequately prepare a thorough report and presentation. In order to gain maximum benefit from this analysis, we have to prepare a detailed report to present to any and all interested parties. The report format is based primarily on style. This may be our own personal style or even a mandated company reporting style. We suggest the following items to be included in the report:

- **Explain the analysis**—Many of our readers may be unfamiliar with the Opportunity Analysis process. Therefore, it is in our best interest to give them a brief overview of what an Opportunity Analysis is and what its goal and benefits are. This way, they will have a clear understanding of what they are reading.
- **Display results**—Provide several charts to represent the data that the analysis uncovered. The classic bar chart demonstrated earlier is certainly a minimal requirement. In addition to supporting graphs, we should provide all the details. This includes any and all worksheets compiled in the analysis.
- **Add something extra**—We can be creative with this information to provide further insight into the facility's needs by determining other areas of improvement other than the "Significant Few." For instance, we could break out the results by subsystem and give a total loss figure for each subsystem. The manager of that area would probably find that information very interesting. We could also show how much the facility spent on particular maintainable items (e.g., components) like bearings or seals. This might be interesting information for the Maintenance Manager. We must use our imagination as to what we think is useful, but by using the querying capabilities of our spreadsheet or database, we can glean any number of interesting insights from this data.
- **Recommend which event(s) to analyze**—We could conceivably have a couple dozen events from which our "Significant Few" list is comprised. We cannot work on all of them at once so we must prioritize which events should be analyzed first. Common sense would dictate going after the most costly event first. On the surface this sounds like a good idea, but in reality we might be better off going after a less significant event that has a lesser degree of complexity to solve. We like to call these events "low hanging

fruit." In other words, go after the events that give the greatest amount of payback with the least amount of effort.

- **Give credit where credit is due**—We must list each and every person who participated in the analysis process. This includes interviewees, support personnel, and the like. If we want to gain their support for future analyses, then we have to gain their confidence by giving them credit for the work they helped to perform. It is also critical to make sure that we feed the results of the analysis back to these people so they can see the final product. We have seen any number of analyses fail because participants were left out of the feedback loop.

That is all there is to performing a thorough Opportunity Analysis. As we mentioned before, this technique is a powerful analysis tool, but it is also an invaluable sales tool in getting people interested in our projects. If we think about it, it appeals to all parties. The people who participated will benefit because it will help eliminate some of their unnecessary work. Management will like it because it clearly demonstrates what the return on investment will be if those events or problems are resolved.

So, if you are struggling with data quality issues in your current data systems and you would still like to determine where to start your RCA process, consider this approach. It will help you learn a great deal about your facility and provide you with the focus to get started with RCA. In the next chapter we will explore methods for utilizing existing data systems to perform a similar type of analysis. This assumes that there is ample data in these systems and that the data is considered to be of good quality for performing an Opportunity Analysis.

6 Asset Performance Management Systems (APMS)

Automating the Opportunity Analysis Process

In Chapter 5 we discussed the manual interview method of collecting event data to determine the candidates for RCA. Now, let's consider automating the process of event data collection. When we talk about automating data collection we are really discussing how to collect event data on a day-to-day basis using modern data collection and analysis tools. When we employ sophisticated data analysis techniques, we actually have the ability to view the data in a way that turns raw data into actionable information.

In this chapter we discuss what is needed to implement a comprehensive event-recording data system. Following are the core activities that need to be established to enable the automated data analysis infrastructure:

Determine your event data elements.
Establish a workflow to collect the data.
Employ a comprehensive data collection system.
Analyze the digital data.

DETERMINING OUR EVENT DATA NEEDS

Once we have satisfactorily determined our performance metrics it is time to determine the data required to accurately report on those metrics. Our data requirements will vary depending on our selection of Key Performance Indicators (KPIs), so we will provide some common data requirements to satisfy the more common metrics.

Since we are focused on collecting event data it is important to repeat what was discussed in the manual method. The definition of event is still critical whether we are performing Opportunity Analysis manually or with an automated collection system. This definition is critical to the process and is typically the place where efforts like these are unsuccessful. As we might imagine, it is very difficult to collect data on something like events when the term has not been fully defined. What might be an event to you might not be considered an event to someone else!

To ensure consistency of data collection, the definition should be clear, concise, and understandable to everyone. Consider the following example to determine if a piece of equipment failed. "Any time a piece of equipment is taken out of service to repair or replace a component." This is an easy definition to understand and takes away a lot of the subjectivity from the data collector. The biggest complaint with most event definitions is that they are too subjective and therefore do not net accuracy or consistency. So follow some good advice and accurately and "simply" define the event for your organization, and then communicate that definition to all the relevant data collectors. Once the definition is in place, formulate an audit process to ensure that the data is being captured in the expected manner.

So what kind of data should be collected when an event occurs? Table 6.1 is a list of common data items that should be collected for any event. This list is by no means complete, but it is a solid base for getting a good event reporting system off the ground. Most asset performance KPIs could be calculated with data in this list.

ESTABLISH A WORKFLOW TO COLLECT THE DATA

We do not want to minimize the difficulty related to collecting event data on a regular basis. The fact is that collecting accurate event data is extremely difficult to do. Event data is different from some other types of data in that it is heavily dependent on human interaction. Take process data, for instance. This data is automatically captured in a disciplined and consistent manner through the use of a Distributive Control System (DCS) or Process Historian. The data is automatically captured with very little human interaction.

Event data, on the other hand, is very dependent on a variety of people collecting data in a uniform way. For instance, one person might view a coupling failure as a pump event while others might associate the coupling with a motor failure. So how do we ensure that the data is compiled in a uniform manner?

First, we need to educate all stakeholders in the need for accurate data collection. In today's busy work environment we are constantly asked to collect an array of data. The problem with this approach is most people have no idea how the data they are being asked to collect is actually used. When this happens we begin to see entries in the Computerized Maintenance Management System (CMMS) stating, "Pump broke." This obviously gives no detail about the events and provides no opportunity to summarize the data for useful decision making. So before we ask anyone to collect data, we need to educate them in how the data will be used to make decisions. Data collectors will be much more motivated to collect the event data if they see that it will be beneficial in helping them to perform their job more easily. It should be a win-win for both the data collector and the data analyst! The biggest sin we can commit is to ask people to collect data and then *never* use it for improving the performance of the operation.

The second step in the process is related to the first in that we need to develop definitions and codes to support the event data collection effort. This means that we need to determine common event codes for our equipment events and then educate our data collectors in the definition of these codes. You might consider ISO-14224

TABLE 6.1
Common Data Items to Collect for Any Event

Data Item	Description	Importance
Functional Location	The functional location is typically a "smart" ID that represents what function takes place at a given location (e.g., pump 01-G-0001 must move liquid X from point A to point B)	High
Asset ID	The Asset ID is usually a randomly generated ID that reflects the asset that serves the functional location. The reason for a separate Asset ID and Functional Location is that assets can move from place to place and functional locations never move. This is the reason we need to identify both event records to distinguish whether the problem is associated with the location or the asset itself	High
Event Date	This is the date that the event was first observed and documented	High
Equipment Category	This is the "high level" equipment that failed (e.g., Rotating Equipment)	High
Equipment Class	This is the actual class of equipment that failed (e.g., pump)	High
Equipment Type	This is the actual type of equipment that failed (e.g., centrifugal)	Medium
Unit or Area	This uniquely identifies where the event took place within the facility (e.g., Unit 01 – Crude Unit)	High
Failed Component	This is the actual component that was identified as causing the asset to lose it ability to serve (e.g., bearing)	High
Event Mode	This is the mode or manner in which the component failed. This is sometimes subjective and may be difficult to determine without proper training and analysis skills (e.g., fatigue or erosion)	High
Model Number	This is the manufacturer model number of the asset that failed	Medium
Material Cost	This is the total maintenance expenditure on materials to rectify the event. This could be company or contractor cost	High
Labor Cost	This is the total maintenance expenditure on labor to rectify the event. This could be company or contractor cost	High
Total Cost	This is the total maintenance expenditure to rectify the event. This could be company or contractor cost	High
Lost Opportunity Cost	This is the business loss associated with not having the assets in service. There is only a loss when an asset fails to perform its intended function and there is no spare asset or capability to make up the loss	High
Other Related Costs	These are costs that might be incurred that do not relate directly to maintenance or lost opportunity (e.g., scrap, disposal, rework, fines, etc.)	High
Out of Service Date/Time	This is the date/time that the equipment was actually taken out of service	High
Maintenance Start Date/Time	This is the date/time that the equipment was actually being worked on by maintenance	Medium
Maintenance End Date/Time	This is the date/time that the equipment was actually finished being worked on by maintenance	Medium
In Service Date/ Time	This is the date/time that the equipment was actually put back into service	Medium

as a guideline for determining your equipment taxonomy and to help you get started with a good code set for documenting events. ISO is the International Organization for Standardization and it has developed a standard approach for the collection and exchange of reliability and maintenance data for equipment. You can find out more about ISO and the 14224 standard on their website at www.iso.org. A great way to train personnel in this is through the use of scenarios. The groups of data collectors are presented with the various codes and their definitions. They are then subjected to a variety of event scenarios to test how they would use the codes in a variety of common situations. When codes are used, it is critical to make the codes specific to the equipment in question and to not overwhelm data collectors with lists of hundreds of options. It is better to use broader code sets that net fewer codes then to overwhelm the data collector with too many choices. With that said, NEVER use codes like Other or Not Applicable. These will quickly become the default catch-all codes and we will not get the data we need.

Last but not least, a comprehensive workflow will need to be established to collect the data described above. Essentially an array of "W" questions needs to be formulated and answered. For instance:

- Who will collect the data?
- What data is important?
- When will the data be collected?
- Where will it be stored?
- Who will verify the data?
- Who will enter the data?

We will answer many of these workflow questions when we discuss data collection systems. As a prelude to this, what many people do is try to use their CMMS as the initial workflow to collect some of the data, and then devise a supplemental workflow to get the remaining data items. This is certainly one method and may be one of the most effective since some key reliability data is being generated through the use of the maintenance system.

EMPLOY A COMPREHENSIVE DATA COLLECTION SYSTEM

To truly automate the Opportunity Analysis process we need to use powerful data collection and analytical tools. Database technology has come to the point where different types of data systems can easily "talk" to each other so that a wide variety of data can be collected, summarized, and analyzed to allow analysts to make informed decisions.

We are going to discuss a method for transferring data from existing Computerized Maintenance Management Systems (CMMSs) into an Asset Performance Management System, or APMS. Before we discuss the interface between CMMS and APMS, let's discuss the role of both of these systems in the operation of a facility.

A CMMS is designed to assist maintenance personnel in the management and execution of work. The main function of this system is to automate the process of getting maintenance tasks completed in the field. This includes things like generating work requests, prioritizing work, planning and scheduling, materials management, and finally the actual execution of the work. However, job closure is typically seen as the least important of the work management process steps and therefore is not done very effectively. Once the work is done, many do not see the value in spending a few extra minutes to document what was done during the job. What should happen is that event history data should be populated, dates updated, and preferably Bills of Materials (BOMs) and task lists updated so that the planning process will be even more efficient for the next job.

Although a CMMS provides a variety of benefits, it was not designed to be an analytical system to provide decision support to Reliability and Maintenance Analysts. It does, however, offer a variety of good data that can be used to perform reliability analysis. For instance, every work order should delineate the asset ID and functional location of the maintenance event, the date the asset came out of service, and the components that were used to repair the asset. There is obviously much more than this, but those items alone can be extremely valuable in determining event probabilities and even optimizing preventive maintenance activities.

An APMS is not designed to handle maintenance work management process and transactions but rather to take that data and a variety of other data to create actionable information in which to improve the overall reliability and availability of the facility. These tools might contain extensive data manipulation tools, statistical analysis tools like Weibull Analysis, Root Cause Analysis (RCA), Risk-Based Inspection (RBI), and many others. We will focus our discussion on how an APMS can be a valuable aid to helping Root Cause Analysts determine the best opportunities for analysis.

So what data can we use from a CMMS that would help determine through an APMS where the best opportunities for analysis might be? Table 6.2 shows some of the common data fields that would be useful in this type of analysis. This data is a solid starting point to performing Opportunity Analysis for Root Cause events. The next step is to transfer this data into an APMS so that the data can be supplemented with additional data about the event and then be "sliced and diced" to determine the opportunities.

In order to make use of this critical data, it must be somewhat easy to find and manipulate. Having worked with Reliability and Maintenance Analysts for many years, we have seen a number of "homegrown" reliability management systems. I am sure that you too can attest to the existence of such systems. For example, what happens when Reliability Engineers cannot seem to acquire the data they need to do their job? They build it themselves! They miraculously go from capable Engineer to Software Developer. I am sure you have seen some of these masterpieces. They

TABLE 6.2
Common Data Fields

Asset ID	Maintenance Start Date/Time
Functional Location	Labor Cost (in-house/contractor)
Manufacturer	Material Cost (in-house/contractor)
Model Number	Total Work Order Cost
Event Date	Unit
Failed Component(s)	Equipment Type

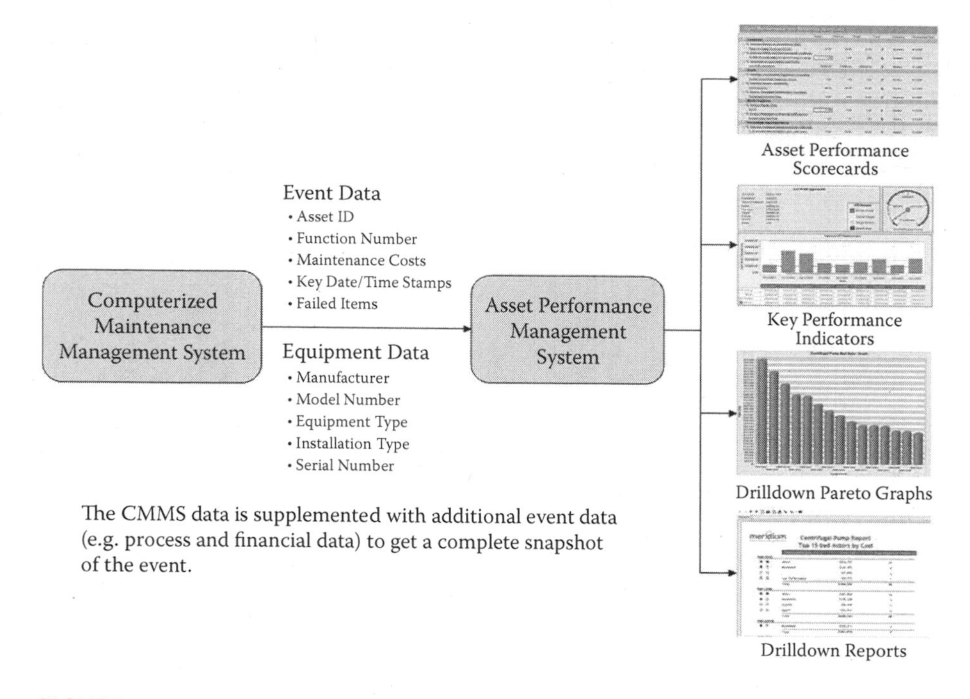

FIGURE 6.1 CMMS/APMS interface.

build them using spreadsheets, desktop databases, or even full-blown development tools. Although these "homegrown" systems serve a valuable purpose for their creators, they have many pitfalls for an organization. For one, the data may or may not be accurate. Since the data is typically collected by a handful of users, it may not truly reflect the overall reality. The data may not be properly event coded, so it becomes extremely difficult to analyze. The main problem with these "homegrown" solutions is that the data is not accessible to all the stakeholders who need it.

An APMS is designed to interface with existing data sources like CMMS, Product Data Management (PDM) systems, process systems, and a variety of others (Figure 6.1). This ensures that the data is accurate and is kept up to date as the interface keeps the system continually in sync. This is critical because it allows the data to be collected once and used for a variety of purposes. An APMS is a secured system, so you know that the data is protected. The most important purpose of an APMS is to provide the value-added analysis tools to turn existing maintenance and reliability data into actionable information.

Let's move on to the area of analyzing your digital data.

ANALYZE THE DIGITAL DATA

The tool of choice to perform Opportunity Analysis is the Pareto chart. Just to recap, a Pareto chart is simply a way to delineate the significant items within a collection. In our case, it will help us determine the few significant issues that represent the majority of the losses within a facility. The Pareto chart can be used on a variety of metrics depending on the need. For instance, some users might simply use maintenance cost as the only measure to determine whether an RCA needs to be initiated. Others might want to compile all the costs associated with an event, namely lost opportunity (e.g., production downtime) costs. Still others might be more interested in Mean-Time-Between-Failure, or MTBF. The assets with the lowest MTBF might be the best candidates for RCA. The advantage of using an automated approach to Opportunity Analysis is that the analyst can look for opportunities using a variety of metrics and techniques.

There are currently some powerful technologies to view and analyze data. One of the best for performing Opportunity Analysis via Pareto charts is a technology called On-Line Analytical Processing, or OLAP. This technology allows users to view data with a variety of dimensions and measures. For instance, suppose you wanted to know which unit within your plant was responsible for the greatest maintenance expenditures? Once you know that, the next obvious question might be which pieces of equipment were most responsible for that? To go even deeper, you might want to know what component caused most of that expense? With OLAP tools, you can use powerful drill-down capability to do this type of analysis. Figures 6.2 to 6.4 are a series of charts demonstrating these dynamic Pareto charts.

The use of OLAP makes sophisticated data mining easy for end users. It allows users to see what they want to see in the form that is the most useful for them. Although OLAP is an incredible tool for dynamic Opportunity Analyses, other tools might be useful as well. Some users might like to see the data presented in a particular format. For instance, suppose there is a corporate reporting standard to which you must adhere. If this was the case, the use of preformatted reports might make the most sense. Reports are useful for presenting predetermined metrics that are

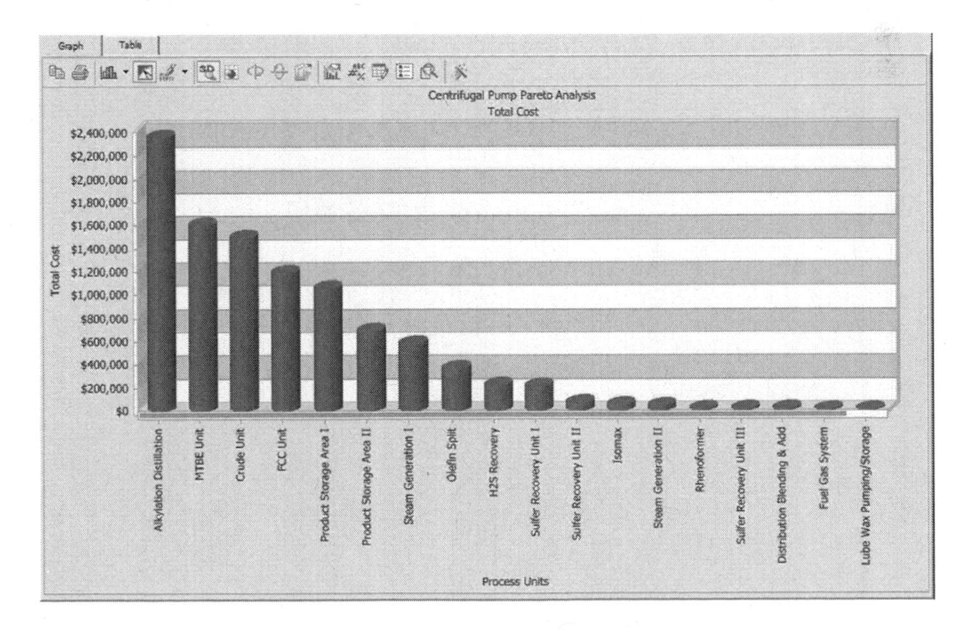

FIGURE 6.2 Step 1—Determine which unit has the highest maintenance cost.

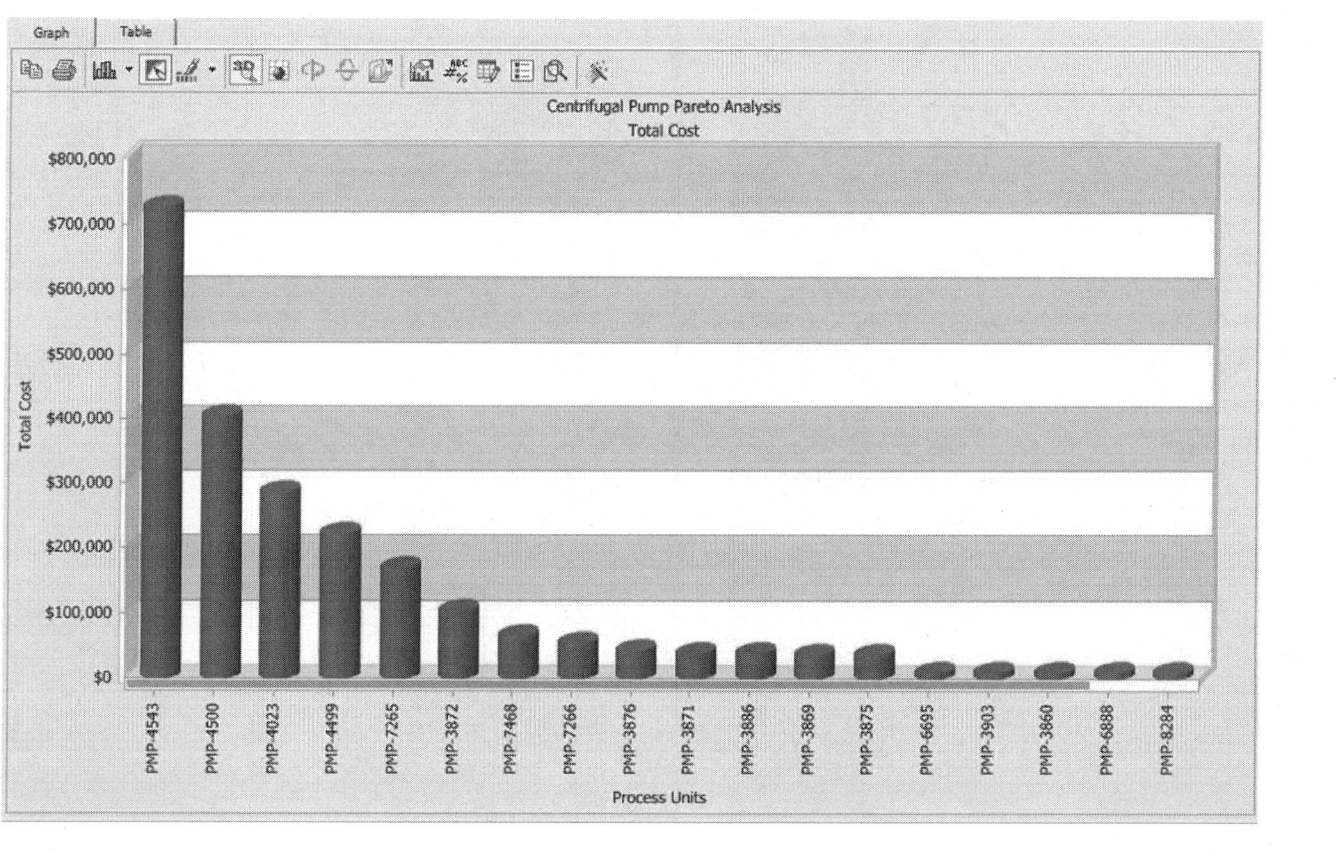

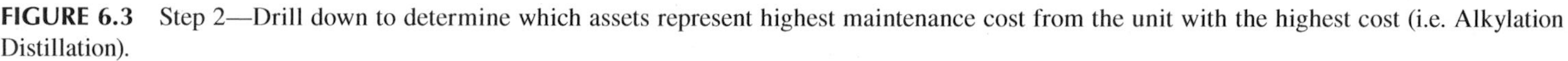

FIGURE 6.3 Step 2—Drill down to determine which assets represent highest maintenance cost from the unit with the highest cost (i.e. Alkylation Distillation).

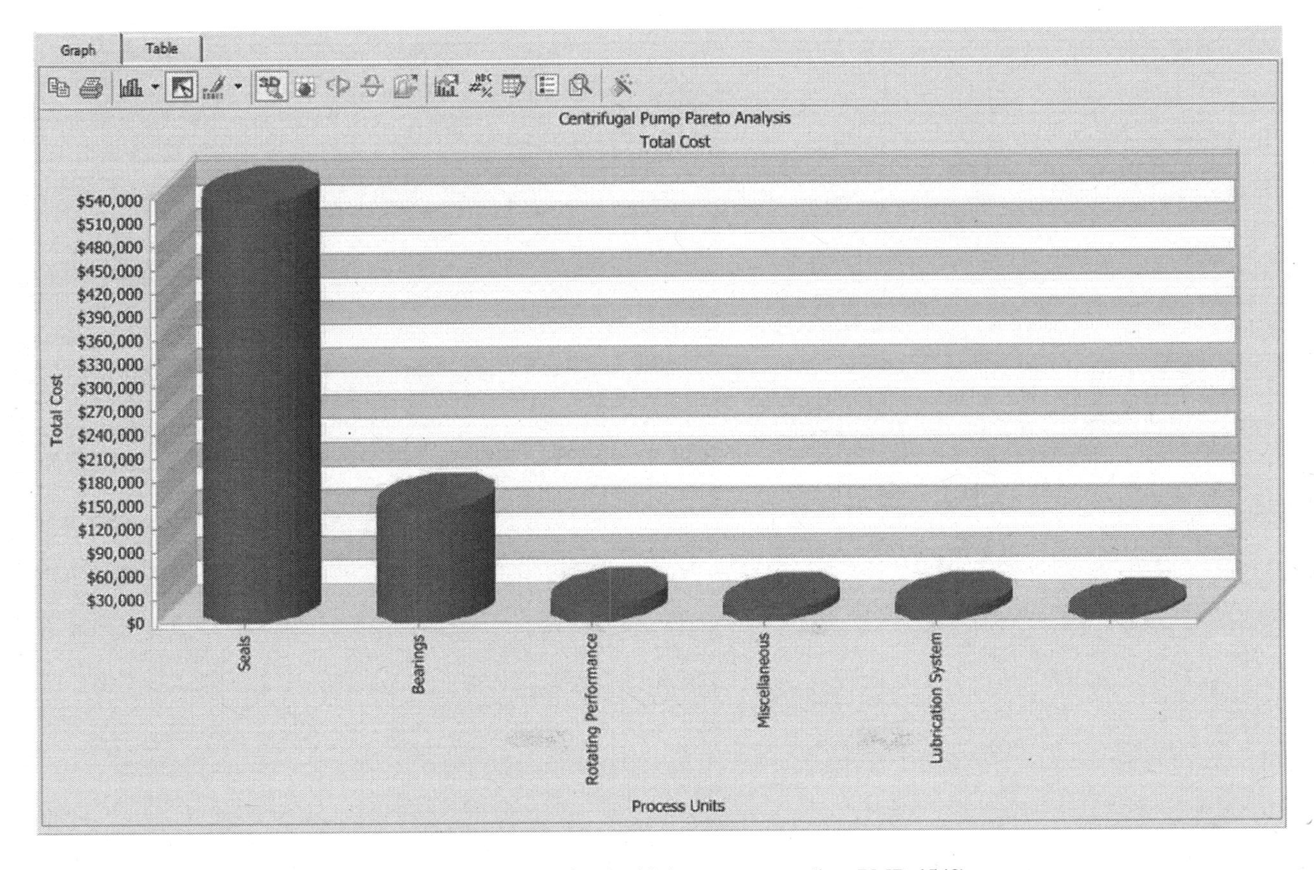

FIGURE 6.4 Step 3—Drill down to determine the components for the highest asset cost (i.e., PMP-4543).

updated every time the particular report is run. Figure 6.5 is an example of a pump event count and maintenance cost report.

To allow for complete flexibility for data analysis, an APMS would provide a comprehensive tool to perform ad hoc queries. A query is simply a way to extract the information we need from the database. This is commonly done using the structured query language, or SQL. SQL is the syntax or language needed to get the relevant data from the database. SQL is not something with which most analysts are intimately familiar. So the APMS must provide a highly flexible query tool that does not require the end user to know anything about SQL. Figures 6.6 and 6.7 show an example of a query designed to determine the MTBFs (Mean-Time-Between-Failures) for a collection of pumps.

This only scratches the surface of what can be accomplished when we automate Opportunity Analysis. There are far more sophisticated statistical methods that can be employed. Our advice, however, is to start with the basics and slowly move into more sophisticated methods.

Automating Opportunity Analysis provides the users with a dynamic tool that allows them to look at opportunities in a variety of different ways. As business conditions change, so can the opportunities. The key is to consistently collect the right data on a day-to-day basis.

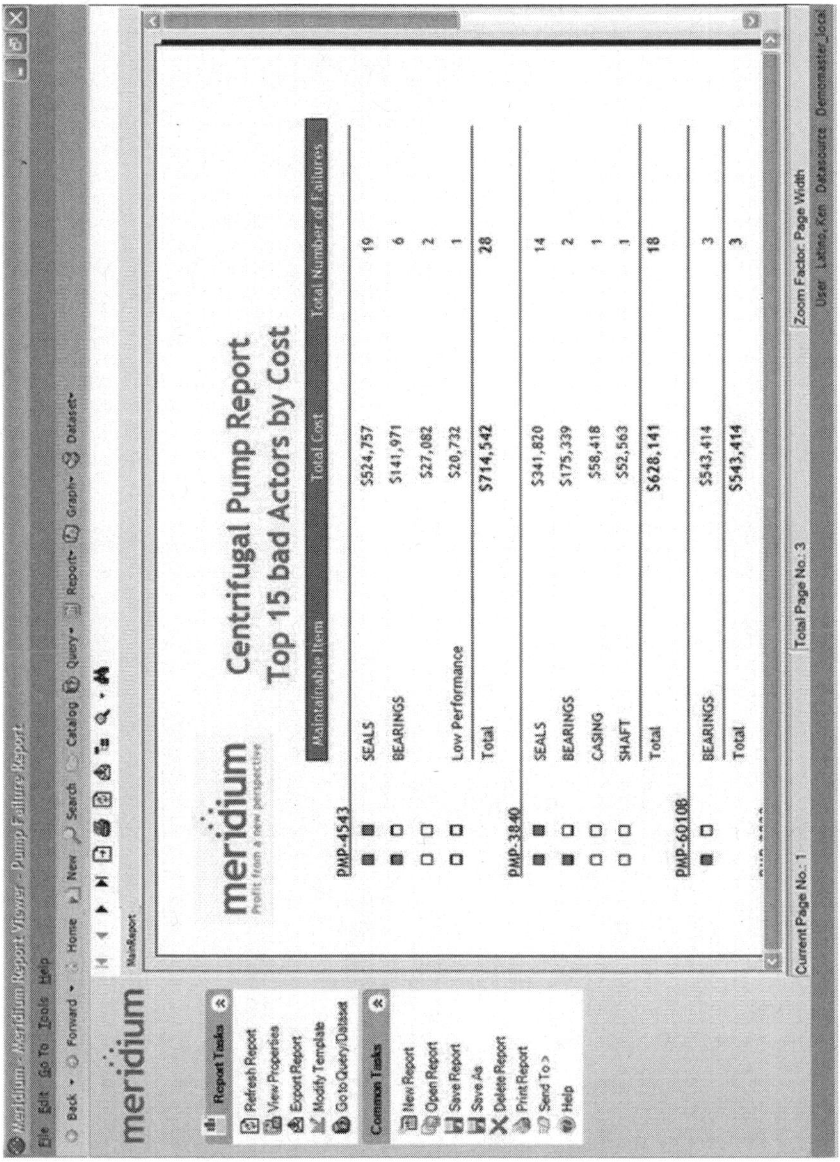

FIGURE 6.5 Sample Pump Event report.

Meridium - View Scorecard - Maintenance and Reliability Scorecard

File Edit Go To Tools Help

Back Forward Home New Search Catalog Query Report Graph Dataset

meridium Metrics

Sitemap: Metrics->Scorecard->Maintenance and Reliability Scorecard

Scorecard: Designer, Manage Privileges, Change Period, Select Columns

Common Tasks: New Scorecard, Open Scorecard, Save Scorecard, Save As, Delete Scorecard, Print Scorecard, Documents, Send To >>, Help

View Maintenance and Reliability Scorecard

	Actual	Previous	Target	Trend	Frequency	Measurement Date
Corporate						
Increase Return on Investment (ROI)						
Return on Capital Employed (ROCE)	21.00	26.00	25.00	↓	Quarterly	6/1/2005
Improve Safety and Environmental Conditions						
Number of overall safety and environmental incidents	1.00	2.00	2.00	↓	Quarterly	6/1/2005
Reduction of Controllable Lost Profits						
Lost Profit Opportunity	150000.00	175000.00	200000.00	↓	Monthly	8/1/2005
Asset						
Minimize Unscheduled Equipment Downtime						
Number of Lost Profit Opportunity Events	5.00	1.00	3.00	↑	Monthly	8/1/2005
Improve System Availability						
Unit Availability	94.10	94.80	97.00	↓	Monthly	8/1/2005
Reduce Scheduled Maintenance Downtime						
Turnaround Downtime Days	13.00	5.00	10.00	↑	Quarterly	6/1/2005
Work Practices						
Reduce Repair Time						
MTTR	11.00	9.00	4.00	↑	Monthly	8/1/2005
Reduce Maintenance Material Inefficiencies						
Average Parts Wait Time	1.20	1.40	1.00	↓	Monthly	8/1/2005
Knowledge and Experience						
Improve Historical Equipment Data Collection						
% of populated required fields in work order history	71.00	76.00	80.00	↓	Monthly	8/1/2005

User Latino, Ken Datasource Demomaster_local

FIGURE 6.6 Sample Maintenance and Reliability scorecard.

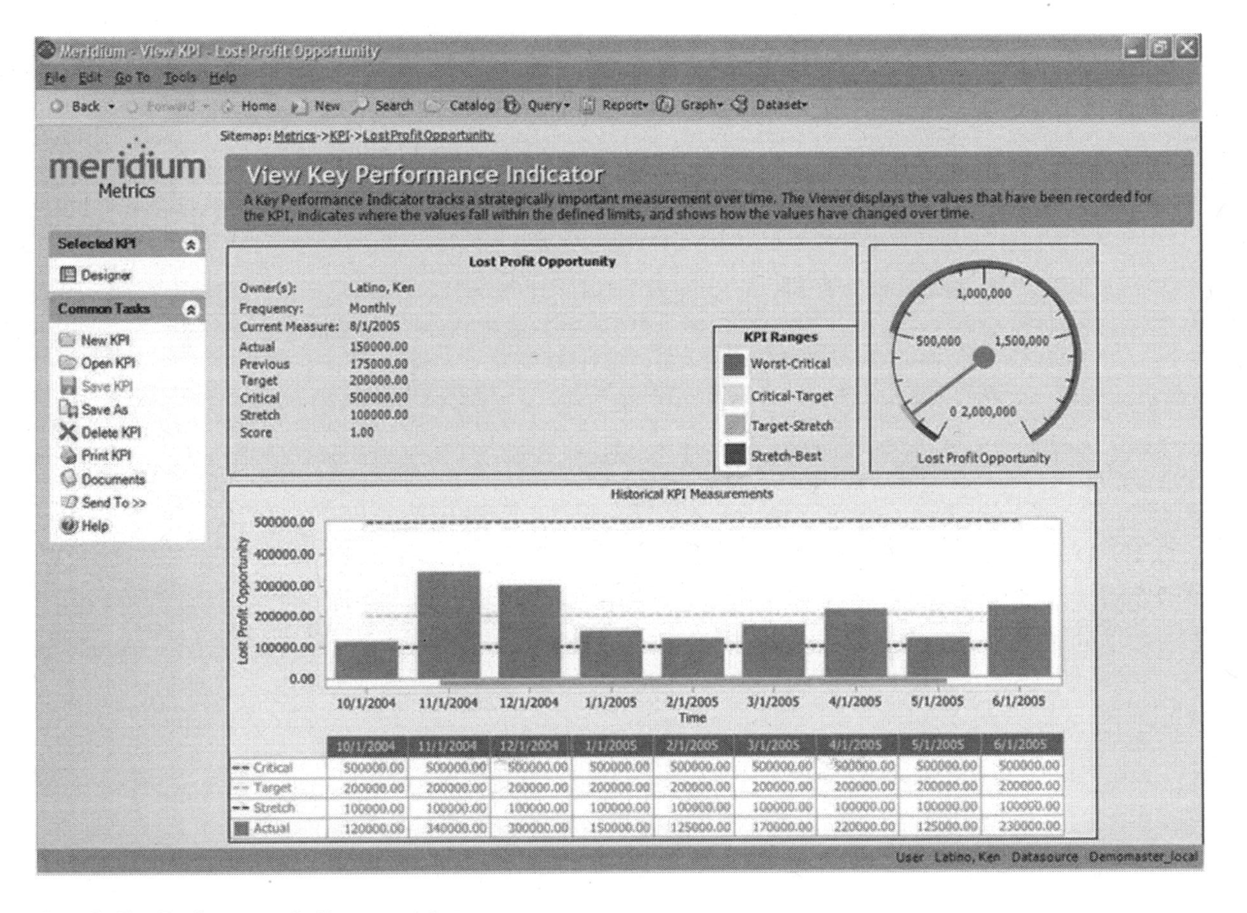

	10/1/2004	11/1/2004	12/1/2004	1/1/2005	2/1/2005	3/1/2005	4/1/2005	5/1/2005	6/1/2005
Critical	500000.00	500000.00	500000.00	500000.00	500000.00	500000.00	500000.00	500000.00	500000.00
Target	200000.00	200000.00	200000.00	200000.00	200000.00	200000.00	200000.00	200000.00	200000.00
Stretch	100000.00	100000.00	100000.00	100000.00	100000.00	100000.00	100000.00	100000.00	100000.00
Actual	120000.00	340000.00	300000.00	150000.00	125000.00	170000.00	220000.00	125000.00	230000.00

FIGURE 6.7 Sample Key Performance Indicator trend.

7 The Role of Human Error in Root Cause Analysis
Understanding Human Behavior

It is important for analysts to have a good understanding of human behavior when investigating an undesirable outcome. Understanding human behavior will help the analyst formulate interview questions. These questions will allow discussion by the employees about subjects that will give insight into the cultural norms present in the organization that could be contributing factors to the event being analyzed.

Each department in any facility has its own way of accomplishing work (some departments' adopted work rules are productive and safe and some are potentially dangerous).

When pressure to meet goals enters the work arena, things can go wrong. Employees will make decisions based on the internal pressures they are feeling and not on the rules and procedures that are in place to protect them. Mechanics will take shortcuts to finish a job on time. Operators will eliminate steps for starting up processes. Supervisors look the other way when they see mistakes being made that may cause delays, and so on.

Let's explore some of the results of decisions made in error. Figure 7.1 shows how using a jackhammer on the same base as a newly installed pump led to the pump failing just hours after startup. Figure 7.2 shows a valve that was installed so close to a support that a pinch point was created, adding to an increased risk of injury. Figure 7.3 shows where equipment was installed on a damaged base with loose anchor bolts, resulting in a failure due to misalignment. Figure 7.4 shows where a pipe was not lined up properly, so an expansion joint was used as a Band-Aid fix, inviting future problems.

Because of decision errors, the analyst should understand what drives human decision error. The top 10 error contributors are

1. Ineffective supervision
2. Lack of an accountability system
3. Distractive environment
    - Low alertness
    - Complacency
4. Work stress/time pressure

5. Overconfidence
6. First-time task management
7. Imprecise communication
8. Vague or incorrect guidance
9. Training deficiencies
10. New technology

FIGURE 7.1 Jackhammer used on base of new pump.

FIGURE 7.2 Pinch point created by equipment layout.

FIGURE 7.3 Loose anchor bolts on equipment base.

When formulating our interview questions, we must keep in mind that the contributors may be in the top-10-contributor list. The contributors to the current RCA should be exposed through our line of questioning whenever possible.

INEFFECTIVE SUPERVISION

Deficiencies in effective supervision are one of the leading contributors to human error. Whether a company has supervisors, team leaders, or peers that lead peers, it does not matter; good leadership is the most important attribute in lowering human error rates. An effective leader is in control of all aspects of the work and in any situation should know exactly what to do. Employees must be confident that the leader's guidance will bring them home safely each day.

The leader must control the flow of work, from the raw materials to the throughput (production, maintenance, purchase, etc.) to the finished goods. The leader must be an effective communicator. The leader is responsible for the production throughput. This means that safety, quality, changeovers, maintenance work, raw materials, and manpower must be sufficiently communicated to produce without incident. The supervisor must know how to coordinate all the departments when any change of normal work direction takes place.

Following is an example of what good supervision should accomplish. One of the responsibilities of a supervisor in a cigarette factory was product changeovers on a line of six cigarette-making machines. A changeover consisted of running one customer order to the end. This involved having the machines completely cleared of old product and materials, cleaned, restocked with new product and materials, and restarted running the new customer order. This should be done in about 1 hour to minimize production losses. The supervisor's ability to use the 4C's of effective supervision (Command, Control, Communication, and Coordination) is a must if

FIGURE 7.4 Pipe misalignment.

the job is to be completed in the allotted time. Each supporting department must have clear and concise communication with the supervisor for the changeover to go smoothly. Each operator, maintenance mechanic, and electrician must know exactly what to do, when to do it, and how to do it. Think about how important it is to have one person orchestrating this entire process and completing it in 1 hour. If more than one supervisor is involved, the chances for success drop because of an additional decision maker, which can complicate and confuse all involved. Bringing everything together when it's supposed to happen is a skill and must be practiced and honed for continuous success.

Leaders must be honest (i.e., walk their talk), trusted by the workforce, respected, and credible for people to willingly follow their lead. Some leaders are picked because they were skilled as a mechanic or operator. Unfortunately, this does not qualify them as good leaders. In many cases this type of promotion produces an average to below-average leader. Look for signs of leadership problems during your investigation. A poor leader can cause significant morale problems, ultimately affecting the quality of human performance.

Supervisors are often the reason for ineffective worker performance, but not necessarily because they are poor supervisors. Often supervisors are picked because they are college graduates or were excellent workers for the company. These attributes, without adequate supervisor training, are not enough, and even with proper supervisor training, some may never be good leaders. The person and the culture have an enormous influence on a supervisor's effectiveness in the field. People who become effective supervisors are usually people who care about employees' success before their own. They generally direct employees to jobs and training that will make them promotable. They are fair and honest in giving feedback to employees, and they provide feedback often so employees know where they stand. The person can be right for the job, but the culture can be flawed, preventing the supervisor from being effective. What this generally means is the culture sets the standard, so when the standard is enforced by the supervisor and the management overturns the supervisor's decision, the supervisor appears to have insignificant authority to management's direct reports. When this is done often, the supervisor can become disengaged, resulting in the loss of a once effective employee. When inadequate supervision comes up in an investigation, do not assume the supervisor is at fault without looking beyond the person for cultural signs that contribute to unsuccessful supervision.

LACK OF AN ACCOUNTABILITY SYSTEM

Accountability is another area with which an effective RCA Analyst must be familiar. Accountability is most often associated with punishment for noncompliance of company rules and regulations. Accountability is often confused with consequences. Consequences are punishment and are only valid if the employees know the rule exists and what the consequences are before breaking the rule. The rule must be consistently reinforced by leadership and must give instant feedback to the employee when a rule is violated. To have a rule that is enforced for one employee and overlooked for another will breed contempt for the supervisor. When the rules are different for each employee and are supposed to be the same, the damage will usually surface in the form of low morale.

Accountability means that a group of employees have agreed on a strategy to meet a particular goal or mission. At the point of agreement the employees become accountable to each other to obtain a higher standard of performance. Some of the key elements to look for when uncovering accountability issues are, first, does an accountability system exist? If it does, is it active? Look for missing data associated with an accepted accountability system such as

- Is there a clear understanding of the goals for which the employees are accountable?
- Are the goal specifics laid out in a clear and understandable manner?
- Are the goals attainable (realistic)?
- Are the results measurable?
- Are the measures results oriented?
- Can the results be tracked?
- Is what we are accountable for considered ethical?

- Is what we accomplish documented?
- Is goal attainment recognized by leadership?

A good accountability system is one in which we help each other by reminding each other of the agreed-upon strategy. If help is needed for another employee to meet the performance criteria agreed upon, it is given (within reason).

Characteristics of a poor accountability system are as follows:

- Based on fear
- Reward levels are the same for all
- Nonperformers still get a reward
- Low performers do not improve
- Punishment is the same for all with no regard for previous performance or circumstances
- Does not inspire employees to perform
- Inadequate teamwork

Accountability issues are often a contributing factor in Root Cause Analysis investigations. We should be aware of their existence and importance when performing interviews.

DISTRACTIVE ENVIRONMENT

A distractive environment means there is an interruption of work about every 5 to 15 minutes. A distractive environment can be areas where noise levels are high, lots of work activity is taking place simultaneously, there is time pressure to hurry work along, there is conflicting management direction, employees are performing multiple tasks, and the like. These types of distractions cause people to lose focus.

Under such conditions distracted employees must move their focus (or concentration) from the work at hand to the distraction. They should focus on the distraction, return to the original work, remember where they left off, and continue.

Distractions often cause procedure steps to be missed, work to be incomplete, tools to be left behind and buttoned up inside equipment, changing to wrong settings, and the like.

The distraction itself can cover an array of things from high noise bursts to leadership continually asking a mechanic, "Is the pump ready yet?" Distraction also comes from excessive phone calls, radio requests, pager calls, etc. A distractive environment can sometimes be reduced by setting rules around whatever the distraction is, but in some cases there is little that can be done. When this is the case, the employees must create their own rules to remember "as-left" conditions. Make note of any distractions uncovered during the investigation. A distractive environment may not always be a direct contributor to a failure, but often it is identified as a contributor.

An example of a distraction being one of the main contributors to an incident is the head-on train collision of Metrolink train 111 and a Union Pacific freight train in Chatsworth, CA, on September 12, 2008. The two trains collided under conditions where signals were clear and there were no obstructions. Evidence was collected that indicated the engineer had sent several text messages seconds before the collision. This distraction led to a number of fatalities including the death of the engineer.

Low Alertness and Complacency

Within the distractive environment are low alertness and complacency. Look for signs that would cause the alertness level of an employee to be tested.

The hours worked up to the time of the incident should be checked. If the employees had worked 30 days straight in 12-hour shifts, mental fatigue should be considered as a possible contributor (direct or indirect).

Review any medications that may have been taken prior to the incident (i.e., cold medicine, blood pressure, etc.). There are many cases where human errors occurred when the employee was medicated because of a severe cold or allergy. We may not always be able to attain this type of information. It depends on the type of investigation as well as contractual obligations and healthcare information privacy concerns. We are merely saying it should be considered.

The amount of sleep attained in the last 72 hours should be considered because human fatigue is known to cause poor decisions. There are two types of fatigue that are important for our purposes. The first is fatigue from lack of sleep, which wears the body down. As mentioned previously, it seems that investigations where employees worked 12-hour-plus shifts for 30 days or more without time off have considered fatigue as a contributing factor. The second is mental fatigue, which is directly related to task complexity and time. The more complex the task plus the time to complete the task, the higher the chance of committing a decision error. A certified welder is a good example for mental fatigue decision errors. The welder performs many welds that will be tested for voids. A high amount of attention is required for long periods of time with stress levels elevated due to testing.

Shift work schedules are important because shifts matter. Shift work causes changes in the body's circadian rhythm, which is the body clock for the human being. When the time of the incident is determined, it can be matched with worker shifts and time within the shift. If the incident happened at 2:30 am, the body's circadian rhythm is at a very low alertness level during that time. Knowing this helps the investigation team see if other collected data supports low alertness due to that particular time of day.

Dramatic life changes can affect alertness levels significantly. Life occurrences such as divorce, serious sickness, death, financial challenges, legal troubles, and the like can consume one's concentration on the job. An example is being diagnosed with a serious illness. As soon as a person hears those words, the mind goes into a whirlwind of thoughts about past, present, and future. It would be extremely difficult to perform a task from beginning to end and be focused. In any case the probability of committing an error is increased under such emotional conditions.

A significant, upsetting argument may also cause alertness issues. If an employee has a significant argument with a spouse or child at home, it can affect his or her alertness level at work. A significant argument with a coworker can also have the same effect.

Complacency can be identified from a number of factors as well. Consider the following.

The number of years performing the same job can be a factor in an investigation. When an employee has been doing the same job for 12-plus years the employee is considered to be in an overconfident state, which can lead to complacency. Think about overconfidence like driving a car. Let's say a person drives to work every day taking the same route. The person has been doing this for 12 years and sometimes seems to go into a trance to the point that the person does not remember the ride home, only the arrival. This is a complacent state of mind and under normal, low stress conditions the arrival home will happen without incident. But when unforeseen occurrences take place (like a deer jumps in front of the car) the driver may not be able to take corrective actions fast enough to avoid an accident. In a complacent state of mind, if the unexpected occurs, stress is added quickly and the person must catch up fast and react. Sometimes those catch-up seconds are not enough to avoid an undesired outcome.

The job's complexity (high or low) level must also be considered. Low stress, monotonous jobs can be a source of complacency. High-complexity jobs require more attention and actually can lower the possibility of error under certain conditions (like shorter time frame for task completion or proper breaks to keep focus). The importance of low and high stress tasks to the investigation is about uncovering too low of a stress level for mental complacency mistakes and too high a stress level without proper breaks for inadequate alertness issues.

The investigation team should consider how often an individual performs a job (skill level). We will define "skill" as performing a task 100 total times and about once each week. Once again the "driving a car" example can apply. Most people have driven a car 100 times and most drive nearly every day and are considered skilled. There are three levels of human error considered with each investigation: skill-based, rule-based, and knowledge-based errors.

These levels were developed by renowned researcher of cognitive systems Jens Rasmussen. Basically what these levels mean is if we commit a skill-based error, it is not because we don't have the skill to do the task; it is more likely that we had a lapse in memory (some type of a slip). It was not done intentionally; it just happened. We did not intend for the outcome to occur. A skill-based error is like applying the correct torque for 9 out of 10 bolts and forgetting to torque the tenth bolt altogether (another common juncture for error is at the end of a planned task).

Rule-based errors, simply stated, means the person maintains the skilled status but for some reason chooses to do or not do something. The person chooses to break a rule and when the rule is broken there is an undesired outcome or failure. Let's look at another car example where a skilled driver decides to run a light as it is turning red. The driver knows he or she is breaking a rule but still does it and is involved in an accident as a result of the action. There are many reasons why people break rules: they are in a hurry to finish a dirty job, it's close to the end of the shift and they want to go home, they want to keep the boss off their back, it allows them to start the process up faster, and the like.

Knowledge-based errors occur because the person does not have the internal knowledge or experience to perform the task at hand. Returning to a car example, a newly trained driver with little "on the road" experience has a higher risk of committing an error than a skilled driver. A work environment example could be the classification of diesel mechanic. It doesn't mean the mechanic has work experience on all types of diesel engines. A Fairbanks Morse®* engine is quite different from a PAXMAN®† engine, but they both are diesel engines. If a PAXMAN®-trained diesel mechanic is asked to perform work on a Fairbanks Morse® diesel engine on which the mechanic is not formally trained, the only knowledge the mechanic has to pull from is his or her PAXMAN® engine experience. This increases the probability of human error and is the reason analysts want to know about training records. The same is relevant when someone who was once skilled but has not performed the task in years is more likely to commit a knowledge-based error.

The reason for looking into refresher training is because the length of time between refresher trainings can be crucial to understanding what the individual's thinking might have been at the time of an incident. An example of why this matters could be described during a planned job to replace a track on a large shovel. The crane making the lift failed by completely separating from the turret base. The rigging was one of the items questioned because the track was rigged with a chain on only one side. This, according to rigging experts, is not the proper rigging technique for this type of lift. The crane operator and the rigger were both experienced and had been doing these types of jobs regularly in excess of 20-plus years each. When questioned about qualifications, it was revealed they were trained during their first years of service and there had been no refresher training since. This matters to the analyst because it opens other avenues of concern such as the crane operator's refresher training, the maintenance worker's refresher training, etc. Human beings over time will always find ways to perform jobs faster and easier. What likely happened is the rigger, for one reason or another, probably had to make the lift with rigging only on the one side, which worked and was maybe even easier and nothing went wrong. The next time he tried it the same way, and once again things went just fine. By the third time there is enough confidence in the new way that there is little thought about the method. It has become the new way (habit) of making this lift. Without refresher training, bad habits become unwritten rules of operation. The problem is that without evaluation, we could be setting ourselves up for disaster. There is a saying in our industry that states, "Violation + Human Error = Disaster." There is a lot of truth to this because most incident investigations uncover violations from procedure. The investigator must learn what reasonable "Mean-Time-Between-Refresher-Training (MTBRT)" is and discern if current intervals are within guidelines.

Using a checklist is a good way to reduce work stress and make sure important tasks are completed. Many workers in the industrial/manufacturing world for some

* Fairbanks Morse is a registered trademark of Enpro Industries Company.
† Paxman is a registered trademark of the GEC Althsom Company.

reason see checklists as an insult to their integrity and do not want to be bothered with filling them out. Airline pilots fill out checklists with every flight and we would feel very uncomfortable if they did not. Pilots are well-trained educated people and yet the checklist does not seem to insult them. However, some operators, mechanics, carpenters, etc., are not willing to use checklists without asking themselves, "Why me?" Because checklists are perceived this way, industrial analysts look into whether a checklist was required and completed and/or whether a checklist was not required and should have been. Checklists are given a lower priority when time pressure is present. There have been a number of incidents where final walk-down checklists were supposed to be performed and were not because the perception was that there was not enough time to perform the task. As a result, some tasks were not completed and an unexpected failure occurred. Checklists are a reflection of the most important parts of a procedure and should be used by skilled personnel to minimize committing a skill-based error, and nonskilled personnel should adhere to the full procedure and avoid a knowledge-based error. The use of either a checklist or full procedure will reduce human error, which increases performance and Reliability. The added bonus is that work stress is relieved for the individual. When the task is checked, we know it is either complete or someone has willfully violated policy and checked it without actually performing the task.

WORK STRESS/TIME PRESSURE

Work stress and time pressure can also cause people to make poor decisions. Poor decisions under work stress and time pressure are usually in the form of risk avoidance, less consultation with others, considering fewer options, and proceeding with poor information (less qualifying, validating, and verification of "at hand" information).

People will take enormous risks in the name of production: leaving work on time, staying clear of a boss they don't like, getting the job done fast, etc.

As an investigator the first thing to be aware of about work stress and time pressure is whether the pressure is real or implied. It is reported that time pressure is real about 10% of the time, and 90% of the time the pressure is implied or self-inflicted. Work stress is similar but occurs under somewhat different conditions.

WORK STRESS

We will define work stress as situations in which employees may be looked upon in ill favor by their leadership. This perception may cause the employees to have fear about their unknown standing within the organization and therefore affect their judgment and decision making.

Work stress can come from

- Job–employee mismatch (employee does not like the work, not challenged enough, etc.)
- High workload (inflicted by job and/or self-inflicted by overcommitment)
- Underappreciation for employee's work (employee is not sufficiently recognized for contributions)

- Cultural shock (new company takes over, new management, new rules, new culture conflicts with old)
- Politics and perception (employee must play politics he or she does not agree with in order to survive)
- Taking a promotion the person does not really want (because not taking the promotion may be career limiting)
- Business by the book (no work creativity can be utilized)
- Conflicts between a boss and an employee
- Employee layoffs are coming (fear of unknown employment future)

Time Pressure

Time pressure is a primary reason people make critical errors. Time pressure surfaces under many different conditions. The following are some causes of time pressure:

- Finish a customer order on time.
- Micromanagement of a leader for imposed deadlines.
- Meet production quotas.
- Low inventories need replenishment.
- Beat deadline to retain a new customer.
- Meet performance appraisal criteria.
- Competitiveness of individual to beat a deadline.
- Rewards for beating a deadline.
- Career enhancement.

Let's take an example of an engine that had failed unexpectedly to show how time pressure, whether real or perceived, can cause bad decisions. Since the engine failure was unexpected, there was a lot of pressure to get the repairs completed as soon as possible because the equipment was lost during a critical production run. This ended up halting the completion of the order. When the engine repair was completed, the company's repair process called for the engine to be flushed according to a flush procedure for that type of engine. When the flush was completed according to procedure, it was inspected. The employee conducting the inspection informed management that some debris was still visually present. The management, feeling this would further delay the completion of the order, decided the flush met the procedure requirements and opted to put the engine back into service even though debris was noted in the inspection. The engine was cleared and turned over to operations, which in turn started the run-in procedure. As the engine reached full power, the engine catastrophically failed and was again out of service. In the past, if debris were found after the initial three flushes, it was flushed until the visual inspection was clear of debris. In this particular incident, the normal process was compromised because of the pressure of getting the asset back into service, which caused a conflict with normal protocol. This in turn caused bad decisions to be made.

Understanding the kinds of time pressures present can help the analyst build a number of possible scenarios that can be tested against the collected incident data for possible fits. Any time priorities are in conflict, the potential for poor decisions is possible: production first or safety first, turning a report in on time or being late, order completed on time or quality first, perform maintenance or limp along to next shutdown, etc.

These are the kind of events that should flag the investigator's interest and be expanded upon for understanding. It may also spark additional data to be collected for verification (connecting the dots).

OVERCONFIDENCE

Overconfidence can lead to complacency, unnecessary risk taking, and tunnel vision. Some of the signs to look for are long, consistent runs because the employees are, for the most part, in automatic mode. Automatic mode over time causes many employees to lower their guard, making them complacent and vulnerable to an incident. Now let's flip the coin and look for lots of downtime and comments like, "I work on this so much I could do the job with my eyes closed." Let's look at a different scenario but the same complacent result. There may also be inferred experience noted, especially in an area with newer employees. You may hear things like, "I did similar work at the last place I was employed," inferring that the two jobs are the same. Inferences like this should flag the investigator to dig deeper.

Overconfidence increases with years of experience doing the same job. A rule of thumb for judging overconfidence is if an employee involved in an incident has worked at the same job for 12 years or more. If this is the case, there is reason to suspect that overconfidence played a role in the incident. What happens is the employee has performed the same type of tasks for so long that the employee fails to properly evaluate the risks associated with a particular course of action. Rather than check on something the employee may have forgotten to do, the employee assumes he or she must have taken care of it.

As the investigator, the years of experience of the employees involved in the incident should flag you to look deeper into reasons why overconfidence could have played a role.

FIRST-TIME TASK MANAGEMENT

The first time a task is performed is the opposite of overconfidence. The employee is new at the job and has little experience. First-time task performance is a critical juncture for human error and the way the supervisor handles the learning curve makes a difference. The employee must perform the new tasks to become experienced, but with guidance from a seasoned employee or, better yet, an instructor. The way things are today with reduced resources and training budgets being cut, plant sites have had to "get by" using the buddy system for training. The buddy system simply passes the task education from an experienced employee to an inexperienced employee. The problems to look for are bad habits (shortcuts) that have been passed to new employees as a result of budget cuts.

IMPRECISE COMMUNICATION

We are all accustomed to communicating on a daily basis, and there are currently so many ways to communicate that the chances of a misunderstanding has actually risen with the addition of technology. We have technologies and other means of communication such as

- Radios
- Pagers
- E-mail
- Texting
- Written communication
- Tweeting (using Twitter)
- Sign language
- Signage
- Cell phones

And we are sure that there are other methods of communication such as drums or smoke signals that we have not identified. Because of all the technologies available, critical communications must be clearly laid out in job plans as well as during normal operations. Not only is it important to be clear about the method of communication but also who is responsible for communication when performing tasks. Communication errors have been deemed as the cause of many accidents in the public, military, and industrial sectors. Aircrafts taking off from and landing on incorrect runways, losing a Mars orbiter in space, battles lost in wars, and deaths in industrial accidents.

An industrial example that comes to mind occurred in a power generation plant that was in the process of starting up after an outage. Operations suspected that there may have been water left behind in the generator after the contractors buttoned everything up. The engineers were aware of the situation as well as the maintenance crew and both were checking the drain valve making sure there was no water coming from the generator. The operator was busy doing other tasks associated with preparing for the start-up and was supposed to perform the final check (walk down) before starting the equipment. With all the additional help, the operator assumed the valve was opened and it wasn't. The generator did have water in it and went to ground during start-up, bringing the entire unit down again for additional repairs. This is a classic case of human error because what was considered normal was interrupted by the suspicion of water being in the generator. Because of the interruptions the operator failed to complete the walk down, which most likely would have been completed and the mistake caught. The proper systems were in place but because of a change in work direction/pattern, communication failed and the system did not function as it should have. Each of the groups of individuals (engineers and maintenance crew) were trying to be of service and doing what was considered the right thing, but the communication collapsed and as a result an undesirable event occurred.

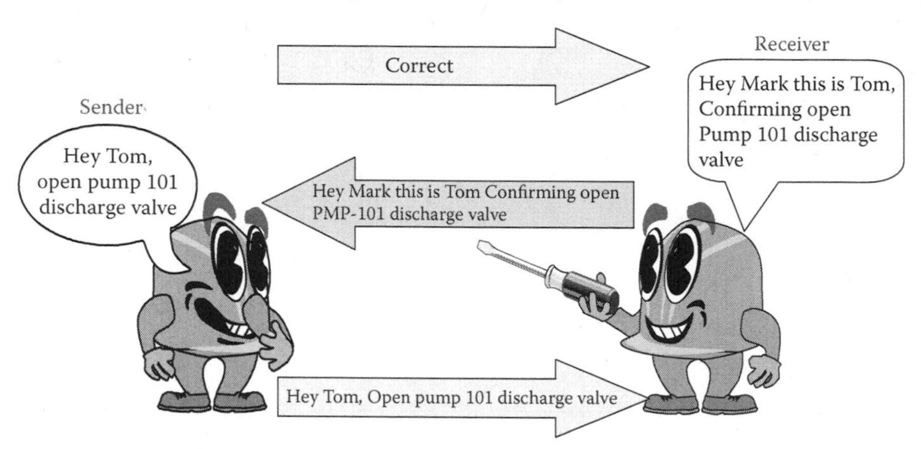

FIGURE 7.5 The communication cycle.

Communication between individuals means that a message is sent by one individual with a particular intent and it is received by another individual and is processed into a meaning or completed cycle (Figure 7.5). Problems arise when the intent and the received meaning do not match up or the communication is assumed as understood by the receiver without any clarification when in reality it is not understood. A communication cycle consists of a thought (information exists in the mind of the sender), encoding (the thought is relayed to the receiver in words), and decoding (the receiver translates the words into concepts or information he or she understands).

To help this process, we must understand how people process information or decode a communication. We are sent a message by a sender that assumes the message is clear, concise, and accurate to the receiver, but it may not be, based on several influencing factors:

- Input data
- Subjective factors
- Human physical properties
- Physiological functions
- Psychological functions

The input data may not be accurate because it needed interpretation, it was missing critical information, or it was misleading. If any of these issues occur, processing will be affected and the output or action may be erroneous.

Subjective factors are a person's ethics, attitude, and social climate. These are essentially a person's values or belief system. It is somewhat static but plays a role in the processing of incoming data. If the data goes against the person's beliefs, it will not be processed into information properly and most likely will be rejected as a task that the person would perform and result in inaction.

Human physical properties are a person's natural given abilities or disabilities and again are more static than not. As people age, their abilities change and may affect communication. An example of human physical constraints affecting decoding or processing is when a transmission is interrupted incorrectly because of a hearing impairment. An example is when someone says, "I thought you said at 1500 hours," when 1100 hours was really stated. Other factors that can affect decoding are when distractions and multitasking are added along with a hearing impairment.

Physiological functions that affect decoding or interpretation are illness, sleep deprivation, fatigue, alcohol, drugs, and the like. When experiencing conditions such as these, a transmission can easily be received incorrectly. Human physiological conditions are dynamic in nature as they can change from one day to the next and will influence information decoding or processing. If someone is ill and possibly taking medication, it is easy to understand how the person could misinterpret information and make a wrong action.

Psychological functions are the most dynamic in that they can change a person's ability to interpret information in an instant. Psychological functions are things like frustration, fear, anger, confusion, and anxiety. When someone is in an emotional state of mind like confusion, the likelihood of committing a wrong action is elevated. If you have ever been through a divorce or some other traumatic experience you know how hard it is to focus at work, and the likelihood of a wrong action is high under such conditions.

Another form of communication that can be problematic is signage within a familiar area. Signs are meant to alert humans to dangers so they can take appropriate measures to avoid injury. However, when we become accustomed to seeing signs on a daily basis the signs are no longer interrupted with "noise" or too much detail. An injury or even a death can result because the sign was seemingly ignored. This phenomenon is called graceful degradation or a form of complacency. Over time we very slowly (gracefully) become unaware that the signs even exist. That's when the danger becomes latent but not active. At this point the danger can become active at any time, causing an event.

Signage can be displayed to be more effective in a complacent state of mind. Because we as humans scan information so quickly, much of our interpretation is based on filling in the blanks. In other words, we scan the information and assume the outcome based on past experience. A good example is being able to read sentences that are horribly misspelled like, "O lny srmat poelpe cn raed tihis" (should read as only smart people can read this). We can surmise what the sentence is saying as long as the first and last letters are correct. Now let's take what we know about humans and scanning and apply it to signage—the rules are the same. Many signs in manufacturing facilities look much like the signs in Figure 7.6. Of course, this is dependent on the type of facility and the environment to which the signs are exposed. The signs obviously are there for a reason, but as you can see there is very little print left on this sign to warn a person of danger. The first thing a person scans is Caution and Safety. Many signs start their message with these words, which means they may not be providing sufficient information to avoid an incident when the individual only scans the area.

Now let's look at a sign provided by a vendor. Figure 7.7 shows the word "DANGER" at the top of the sign and "HARD HAT AREA" in large letters at the bottom. If we

FIGURE 7.6 Typical signage in a manufacturing plant.

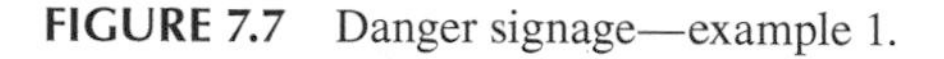

FIGURE 7.7 Danger signage—example 1.

are only scanning words, the person will not get the action message and may become complacent about wearing a hard hat. More than likely people will not comprehend the large words "Hard Hat Area" below the word "Danger."

People may not see words completely, but we do see pictures in a more complete sense. Figure 7.8 shows a sign we created using the rules of scanning. The first thing scanned is a picture of a person wearing a hard hat, which is the message I want conveyed to an employee. The words below the picture state, "Hard Hat Required In This Area," which may be scanned but most likely will not be interpreted. The word "DANGER" is still visible at the bottom, but the action message that provides the data to stay safe is interpreted with the picture immediately at the front or with the first word as seen in Figure 7.9.

When individuals have worked in an area for many years, the picture of the hard hat should be enough of a queue that a hard hat is required. We know the Occupational Safety and Health Administration (OSHA) has some approved common formats with the words Caution, Danger, Safety Alert, and possibly others at the top of the sign. The point is when we want to communicate a message it may not always fit into a preprogrammed format, and in those instances the format may have

FIGURE 7.8 Danger signage—example 2.

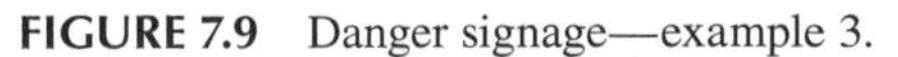

FIGURE 7.9 Danger signage—example 3.

to be reoriented, possibly bringing the word "Danger" to the bottom instead of the top to compensate for the human tendency to scan information.

Body language is also an important part of communication and, when coupled with verbal exchange, makes the clarity of the meaning even stronger. We will explore this further when we discuss interviews during the course of an RCA. About 55% of human communication is done through body language, so it has a significant role in communicating. Body language can tell the sender if the person receiving the message is confused without the person saying a word. It can also give indications of anxiety, boredom, anger, and a host of other emotions that help the sender to determine clarity. About 35% of communication is through tone of voice, which can be useful when communicating with radios, cell phones, and the like. Since the visual patterns cannot be determined, the tone of voice will help with clarity. Like reading body language to determine emotion, humans can often also hear confusion, anxiety, and boredom, as well as many other emotions in one's tone. The spoken word alone is the least remembered form of communication. When words are spoken there are no visuals for people to attach in their memory, and therefore the message is lost usually within 15 to 20 seconds. When words are spoken in a monotone manner, there are no changes in tone or pitch to elaborate on a point and again the message can be lost.

The most ineffective communication can be through e-mail, as emotion is difficult to assess with e-mail. However, e-mail does have advantages in that the sender and receiver can send qualifying questions back and forth rather quickly until common understanding is achieved. E-mail many times is taken the wrong way because people will try to read emotion into the message, which in many cases is misconstrued.

Other communication is made through colors used to signal actions. We are very stereotypical in nature and we expect certain communication to be received in certain ways. Generally, we expect

- A switch to be on when it is in the upward direction
- A knob that is off to be in the counterclockwise direction
- Colors that are light blue to mean up (like the sky) and light in weight
- Dark colors such as brown to mean down (like the ground) and heavy in weight
- A far distance to be bluish (like heat in the distance)
- Red to be hot (as in hot temperature) as well as signify danger
- Red flashing lights to indicate an emergency
- Blue to be cool (as in cool temperature)
- Small boxes that are on top of a pile to be light in weight and large boxes that are at the bottom to be heavy in weight

An example of human expectation involves mugs. Have you ever been served a drink in a mug that resembled glass but when you lifted the mug to take a sip it was light in weight (because the mug was plastic)? The expectation was heavy like you would expect with glass and you most likely used a little more force than you would have if the expectation was plastic. Another example involves adjustable desk chairs. If you have ever gone to sit, expecting the height of the chair to be at a certain level and it was lower, then you know what we mean.

The point is management should insist that employees follow good communication techniques including understanding the importance of body language and awareness of human expectations when designing work environments. Solid communication techniques will help avoid higher error rates in the workplace.

VAGUE OR INCORRECT GUIDANCE

Guidance usually comes in two forms: from written documents like procedures/manuals and from individuals like supervision and peers. Let's start with written guidance and talk about what really happens with procedures, manuals, and the like.

Procedure and manual changes are a common recommendation in incident investigations. When during an investigation it is discovered that if 15 additional steps had been in the procedure the incident could have been avoided, then the steps are then added as a result of the recommendation. It is accepted by most organizations the 15 additional steps will avoid a future reoccurrence. Our own experience tells us this kind of thinking is a myth. We know that skilled, experienced employees seldom use procedures/manuals in such a way as to follow each step, line by line, until a task is complete; it's just not the way we operate as humans. Adding 15 additional steps as a correction does not eliminate the recurrence of the incident; it merely satisfies a site and/or regulatory requirement for completing an investigative cycle. Adding steps to a procedure that is already not being used will do nothing except add more steps to a procedure that is not being used.

Even if by some outside chance employees follow the procedure, adding more steps to remember will only add clutter and questions about the additional steps in the procedure or manual. The paradigm being encouraged is that by adding more steps to a procedure/manual we will reduce future failures. Often after the implementation of such additional procedural steps, the incident will recur. This is one of the reasons so many incidents have happened that were thought eliminated because they were analyzed and recommendations were in place. The quality (depth) of the investigative process matters when incident elimination is sought.

Make no mistake about where we stand on procedures and manuals—they are necessary in the workplace but they must be crafted to fit into the natural behavior of human beings in order to be of value. Let's examine for whom procedures and manuals are written. Are procedures written for lawyers, skilled craftspeople, skilled operators, new craftspeople, or new operators? Should the procedures and manuals be written at a college level, high school level, or eighth grade level? It is essential to know who the target audience is for the written guidance if the mission is to have successful task outcomes.

Once it is determined who the audience is and the grade level in which the written guidance will be crafted, the content and format of the document can be examined. Some problems with written guidance that have been encountered are

- The procedure/manual refers the user to other written guidance volumes (too complicated).
- There are vague instructions that must be interpreted by the user.
- Words are not simple to understand by the user (wrong grade level).

- Paragraphs run together.
- All capital letters are used (hard to read).
- The procedure/manual is outdated.

When there are references to other procedure and manual volumes, the process becomes complicated to the user. The procedure/manual user will almost always take the path of least resistance and most likely not refer to an additional document. Removing this barrier so all the information is contained in one written document should make it easier and promote use of the document.

Vague instructions that need interpretation can be called written human error traps. Most often they are embedded in procedure steps using works like

- All
- Affect
- Analyze
- Determine
- Check
- Normally
- Sufficiently
- As soon as possible

"All" may be used as, "All testing points shall …" The same action can be stated with more clarity as, "Test points 1, 3, and 6 shall …" Written instructions should be short statements that are clear, specific, and accurate.

Another example is the "As Soon As Possible" (ASAP) statement. It is sometimes stated as, "Notify operations as soon as possible after taking the liquor sample." It can be more clearly stated as, "Notify the lead operator on shift within 10 minutes after the liquor sample is taken." The second statement clears up the vague term of operations by using the lead shift operator (one position as opposed to many). ASAP can have different meanings to all involved; it can range in meaning from immediately to next week depending on who is interpreting the statement. The second statement clears up the vague term of ASAP from many interpretations to one interpretation (within 10 minutes).

Procedures and manuals are most often used by newer employees and should be written in such a way for the new user to follow the written procedure/manual and have a successful outcome.

We have seen very large losses occur because operational parameters had changed with no review and verification of procedures. In one case a beer facility experienced a 3-day production loss because the regular operations person was not there and the person filling in followed the procedure for measuring the amount of carbon dioxide needed to sustain the process while the vendor had their annual turnaround. The procedure no longer reflected the reality, and as a result the area ran out of carbon dioxide and was forced to go down. The operator who was absent knew about this problem and had created a method of measurement that would calculate the need correctly. Unfortunately, the operator was the only person with this information and the procedure for the task was never updated.

Experienced employees usually do not follow procedures because of the frequency of application. Although experienced employees generally do not need to follow a procedure, they should follow a checklist. Even though they may have performed the task numerous times, experienced employees are still susceptible to committing a skill-based error, and because of this, checklists that reflect the key steps of a procedure should be used.

Guidance given to employees by supervisors can sometimes be incorrect and if followed could cause an undesirable incident or event. The guidance given by supervisors that is "what is not said" may send an incorrect message to the employee. An example could be a supervisor who is hurrying to attend a meeting. The supervisor observes an employee up on a ladder that is not tied off and says nothing to the employee about correcting the situation. The unsaid message sent is, "The supervisor is okay with this practice," which is incorrect guidance from supervision.

Other ways supervisors can inject incorrect information is when they do not know where an employee is in the task execution. An example is an overhaul of a piece of equipment that had failed catastrophically. The operations people are nearing the end of the outage and are in the process of flushing the equipment to remove any debris that may have been left behind from the maintenance. The supervisor says, "You have flushed enough. Now get the equipment started." The operations people say they think another flush or two may be needed, but their request is declined. The equipment is started and fails within hours because debris was still in the equipment. The supervisor guided the employees incorrectly by not taking the time to properly assess the situation.

Another area where problems can occur is when employees are trained by other employees (the buddy system) to perform a task. Experienced employees know how to perform their jobs but over time have picked up bad habits that are passed on to the newer employees they train. This is a problem because usually when we as humans are taught to perform a task and adopt the task, we are hard pressed to change it. This means if a person learns to do a task wrong, the person will do it wrong consistently and will most likely revert back when asked to change the method.

An example is when someone is taught to install V-belts using a screwdriver while the equipment is running. Then the person is told to shut down the equipment, lock the equipment out, loosen the driver, replace the belts, realign the equipment, remove the locks, and start the equipment back up. The person already knows he can perform the shortcut because that is what he was taught and the correct way takes way too much time. The chances are when someone is watching, he will perform the task correctly, but when no one is watching, he will perform the shortcut. This is dangerous and can cause harm to both the employee performing the tasks incorrectly and other employees as well.

Employees should be trained to perform job tasks correctly by qualified instructors. This will ensure that the employees can do the job the company is asking them to do safely as well as correctly. Buddy training can come into play after they have received certified Instructor-Led Training (ILT). The new employee should see the dangers of the experienced employee's bad habits because now the new employee knows better. The new employee may even question the experienced employee about the dangers of his or her bad habits. If human traits stand true, the first way the task was learned should be the method to which the employee will revert when in doubt or when bad habits creep in.

TRAINING DEFICIENCIES

Training deficiencies are universally common. It appears that industry is so focused on cost that many important aspects of a healthy work environment are sacrificed to attain a better-looking bottom line for stockholders.

It seems that when sales are slow, one of the first programs to be cut is training. We know that in hard economic times, companies must become lean in their spending activities. But training the employees to perform at the highest possible standard is a necessity, not a luxury. In most cases a company's maintenance of physical assets like machinery will not be sacrificed in low sales cycle times. So why would a company neglect the employees who are expected to run the assets and produce product at a high standard?

When cost-cutting decisions are made, do employers consider their moral responsibility to provide a safe work environment for employees? A safe work environment is not just about providing personal protective equipment and machine guarding—it is also about providing the training for employees to be proficient at doing the job the company is asking them to do. The company should entertain several thoughts about their training program before discussing training cuts:

- Can the employees currently perform the job correctly?
- How do employees know they are performing the job correctly?
- How do employees know the outcome is error-free?
- Do employees know what to do if there is a problem?
- Do employees know what support they need from their supervisors?
- How many past RCA investigations indicated employees were in error?

If employees don't know, they won't do their jobs error-free. When employees know how to perform their jobs, many human error traps are eliminated. A benefit from knowing how to do your job is most apparent when critical decisions must me made and they are made correctly and with certainty. Many times critical decisions are made with uncertainty because the situation has not come up very often and the employees will make the decision based on what they think is correct, not what they know is correct.

A good example of this is the Continental Flight 3407 accident. The flight was operating from Newark, New Jersey, to Buffalo, New York, when it was involved in a deadly accident at about 10:20 pm on February 12, 2010.

The National Transportation Safety Board's (NTSB) main finding of cause was the captain's inappropriate response to the activation of the stick shaker, which led to an aerodynamic stall from which the airplane did not recover. The pilot should have pulled the stick up, and instead he pushed down (which was in error) to avoid the stall. It is very likely that this decision was made because he thought and didn't know what the correct maneuver was. This is where training intervals really matter. Maybe training had been delayed or cut back due to cost cutting—we don't know. We do know that the training issue will be investigated and recommendations will be suggested for improvements.

Stress is reduced when people know what they are supposed to do because they don't have to worry about whether they are making the right decisions. They also know how long tasks are supposed to take, minimizing internal feelings of time pressure. When job pressure is reduced, employees are more likely to take the time to consider more alternatives before taking an action, lowering the probability of decision errors.

With that said we cannot leave out the fact that people learn differently. Some of the ways we learn are

- Formal training
- On-the-job training (OJT)
- Self-teaching
- Consultation

Formal training is always a good idea because you want employees to be trained without shortcuts. You want them to know the best technique to perform a job so the company's assets (People, Process, and Equipment) are protected while still delivering high performance.

On-the-job training (OJT) should not be confused with "buddy training"—they are two very different methods. OJT is ideal after a student has attended formal training. It is in the best interest of both the student and the company for this to take place the very next week after the completion of formal training. It is also recommended for employees learning new information, so they have someone with experience there to observe until the employees have demonstrated they can produce successful outcomes and are comfortable. The reason there should be an observer present is because when people have learned new information they have a high probability of committing a "knowledge-based error." Buddy training, on the other hand, is not recommended without first receiving formal training. Oftentimes companies will replace formal training with buddy training to save costs. Buddy training generally passes too many bad habits on to the trainees.

Some individuals can teach themselves—they can read a book and then repeat what they read with actions and deliver a positive result. This is great if you are dealing with someone you know can do this. Most people don't fall into this category.

Consultation, or perhaps better said as "participation with an expert," can teach individuals new information. One of the best training tools is for employees to actively participate on Root Cause Analysis (RCA) projects; because of the RCA discipline people must drive deep into problems. As individuals drive down they learn how things are supposed to work, which is a real learning experience.

For training to be effective for most of us we must be told, we must be shown, and we must be involved to actually understand.

NEW TECHNOLOGY

New technology is the single greatest advancement in the world that has enabled manufacturers to produce more quality product with a minimal amount of resources. The more that new technology is introduced, the less input that is needed from the human element; technology simply can make critical decisions faster and more accurate than humans.

However, the number of correct and accurate decisions made is totally dependent on the quality of the information being generated to the individual. Most of the companies we have been exposed to have, or are in the process of buying, software systems to track almost anything capable of being tracked in an organization. Although this is generally good business one must realize that software systems, new equipment, and the like must still work in conjunction with humans.

New technology doesn't remove the potential for human error—it changes the potential for human error; it relocates it. With new technology, new things must be learned and as with all change there is a learning curve. A learning curve refers to the mistakes made by individuals as they learn new software, new techniques, and new equipment. They must learn the new way to repair, operate, and report information.

Even though it is possible to throw mass amounts of data at individuals, it doesn't mean the data is being interpreted in way that would lead the individual to an error-free decision.

Often employees are told that the company will be installing new software systems as well as new, upgraded equipment. This is a lot to handle for individuals, primarily because new information must be learned, which is stressful. It leads to questions and concerns like, "Will I be able to learn this new software?" "What will happen to me if I screw up?" and "I am too old to have to learn yet one more new thing."

In most cases where change is involved the first thoughts that go through a person's mind are the negative possible consequences of the change. Fear of the unknown is a natural response to new technology as well as any kind of change. We all know change is constant, but people will resist it with everything they have in order to rationalize why new technology or change is not necessary. When a company is purchased by another company, have you ever heard anyone say it is great news and that they will be given great new opportunities for moving forward? We most likely hear, "Oh, great, the new company's going to come in and chop the company up and lay us off." When the future is unknown, human beings will feel fear, not opportunity.

Keep in mind that humans still interact with technology, and as long as human beings interact with technology they must be considered an important asset that must be nurtured. This is where most companies fall short—they do not prepare employees in a way that slowly reduces the fear of the new and ties the new way to the values and/or beliefs already in place.

New technology and change require patience, and employees often cannot and will not change quickly; they must be brought along slowly so that they can attach the new changes to their existing beliefs. If employees do not accept the new way into their current beliefs and values, the change will most likely be short-lived (more like the program-of-the-month).

New technology starts with design. The designer must look at the entire system, not just the equipment or software. In the case of equipment, designers must consider the operator, who should be able to operate the equipment using the lowest number of unnecessary movements (ergonomics). We once worked in a cigarette manufacturing facility where changes were constant, especially when it came to speed. Back in the day, a cigarette-making line was comprised of six machines that produced nine thousand cigarettes per minute. The packer that made the individual packs produced five hundred packs per minute. The cigarette packs were bundled and fed into a carton, which held 10 packs. Each line ran multiple brands that were fed to a six-lane conveying system, which in turn fed the cartons to a boxer where 60 cartons were boxed and sent to the warehouse. The line was capable of running two brands at the same time because each line had two boxer units. To keep the correct cartons from multiple machines going into the correct box, an electronic lane changing system was used to insure that the correct brand went into the correct box (see Figure 7.10).

There was a fairly consistent problem with cartons becoming jammed, causing the cartons to back up until cleared by the operator (see Figure 7.11). The operator

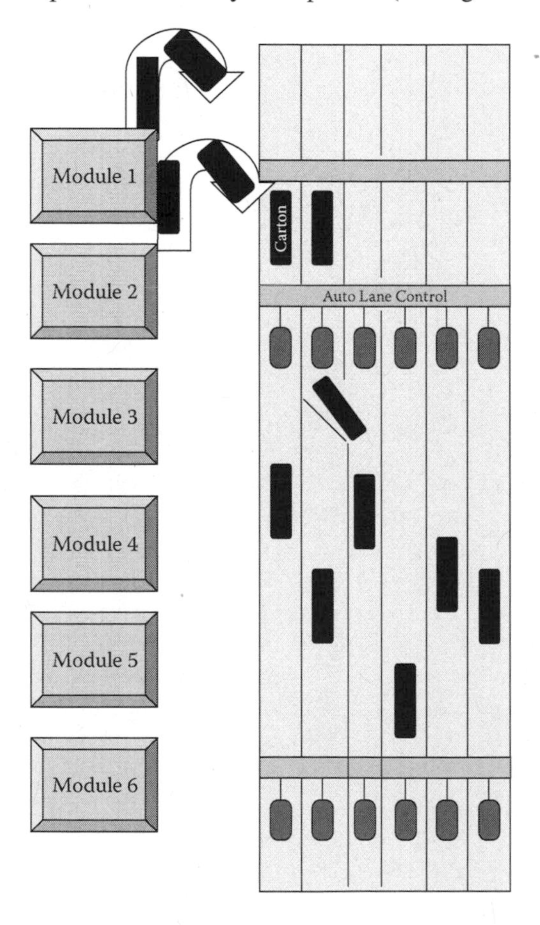

FIGURE 7.10 Electronic lane-changing system.

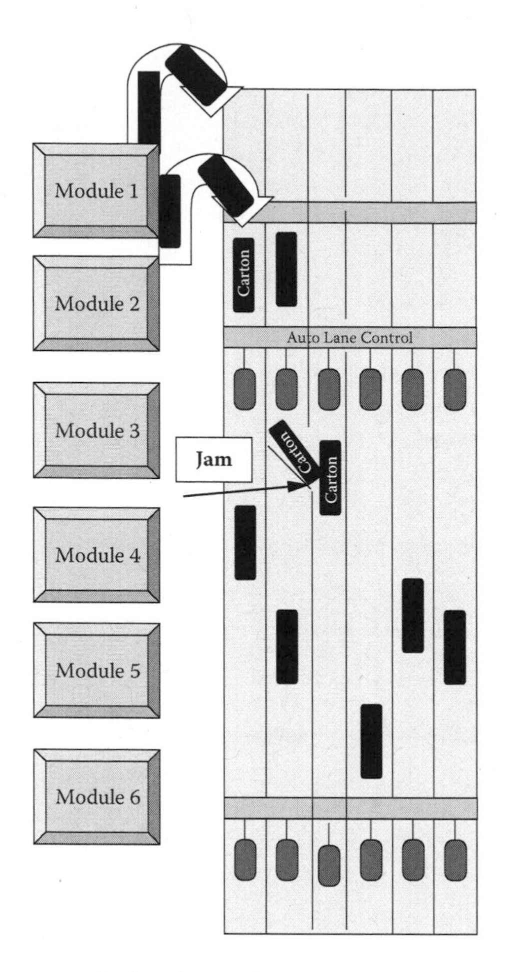

FIGURE 7.11 Operator required to clear up jams.

would go to the conveyor, which was about six or seven feet from the floor, and run his or her hand through the jammed area clearing the jam (there was also a stool at each module's conveyor if needed). This took the operator a few seconds and the equipment was rarely shut down due to backups.

When changes came, they were in the form of faster machines from a different manufacturer with the intention of eliminating the need for one machine from each line. The new design did not involve input from any of the operators or mechanics. The new design had the carton-conveying system about 20 feet from the floor. This was questioned by some because of the jams incurred on the older equipment. The standard answer was, "The technology was now available in the new system to eliminate that problem." To make a long story short, the equipment was installed, the cartons jammed, and production stopped on many occasions. The correction involved a scissor jack the whole area had to share until platforms and staircases could be constructed to allow access to clear carton jams. In all, by not including the people closest to the equipment to participate in the design stage, the overall losses went from a few seconds to clear a lane to as long as an hour. The impact went from rarely shutting down the process to often shutting down the process. This type of trouble can be avoided by taking the time to think through the implementation of change based on the shared values of the workforce along with allowing input from the workforce.

This chapter briefly discussed a number of topics surrounding human error. Keep in mind that each of the topics discussed could be a book—the human element is a never-ending subject of study.

When investigating events always remember that the human being was most likely involved in its unfolding, but most likely it was not intentional. Using what we know about human error and applying it in a proactive manner (managing the human asset) will help avoid future events.

As elaborated in this chapter, the man–machine interface will always be subject to the effect of human error. We will likely never rid ourselves of the risk of human error, but there is much we can do about understanding the conditions that increase the risk of human error and take actions to prevent those conditions from existing. Human beings are not perfect beings, and these imperfections are often transferred in the form of flawed decision making. Poor decisions trigger physical consequences, and the error chain continues until we have to do something about it. In the coming chapters we will discuss in practical terms how to uncover the contribution of human error to undesirable outcomes.

8 Preserving Event Data

THE PROACT® RCA METHODOLOGY

The term "proact" has recently come to mean the opposite of react. This may seem to be in conflict with PROACT's use as a Root Cause Analysis (RCA) tool. Normally when we think of RCA, the phrase "after-the-fact" comes to mind—after, by its nature, an undesirable outcome that must occur in order to spark action. So how can RCA be coined as proactive?

In the last two chapters on Opportunity Analysis (OA), we clearly outlined a process to identify the failures or events, on which it is was actually worth performing RCA. We learned from this prioritization technique that, generally, the highest return-on-investment (ROI) events to analyze are NOT the sporadic incidents, but rather the day-to-day chronic events that continually sap profitability.

RCA tools can be used in a reactive fashion and/or a proactive fashion (Figure 8.1). The RCA analyst will ultimately determine this. When we use RCA only to investigate *incidents* that are defined by regulatory agencies, we are responding to the daily needs of the field. This is strictly reactive. However, if we use the OA tools described previously to prioritize our efforts, we will uncover events that many times are not even recorded in our Computerized Maintenance Management Systems (CMMSs) or the like. This is because such events happen so often that they are no longer anomalies. They are a part of the job. They have been absorbed into the daily routine. By uncovering such events and analyzing them, we are being proactive because unless we look at them, no one else will.

The greatest benefits from performing RCAs will come from the analysis of chronic events, thus using RCA in a proactive manner. We must understand that oftentimes we get sucked into the "paralysis by analysis" trap and end up expending too many resources to attack an issue that is relatively unimportant when considering the big picture. We also at times refer to these as the "political-failures-of-the-day." Trying to do RCA on everything will destroy a company. It is overkill and companies do not have the time or resources to do it effectively.

Understanding the difference between chronic and sporadic events will highlight our awareness to which data collection strategy will be appropriate for the event being analyzed. The key advantage, if there can be one with chronic events, is their frequency of occurrence. This is an advantage because, like a detective stalking a serial killer, the detective is looking for a pattern to the killer's behavior. In this manner, the detective may be able to stakeout where he of she feels the next logical crime will take place and hopefully prevent its occurrence. The same is true for chronic

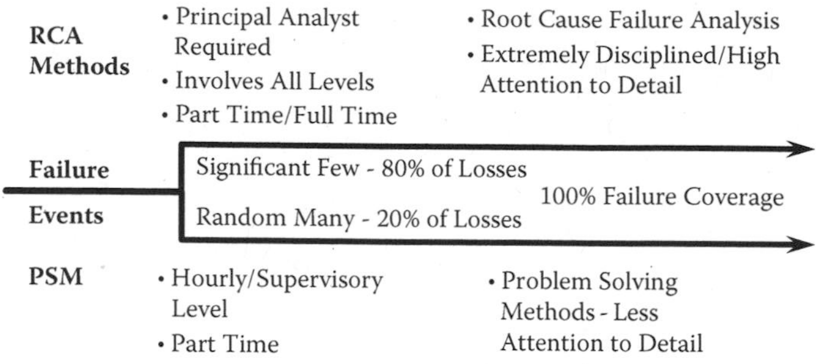

RCA Methods
• Principal Analyst Required
• Involves All Levels
• Part Time/Full Time
• Root Cause Failure Analysis
• Extremely Disciplined/High Attention to Detail

Failure Events
Significant Few - 80% of Losses
Random Many - 20% of Losses
100% Failure Coverage

PSM
• Hourly/Supervisory Level
• Part Time
• Problem Solving Methods - Less Attention to Detail

FIGURE 8.1 The two-track approach to failure avoidance.

events. With chronic events we have in our favor that they will likely happen again within a certain time frame, and we may be able to plan for their recurrence and capture more data at that point in time. We will discuss this more when we discuss verification techniques in the "Analyzing the Data" chapter (Chapter 10).

Conversely, when we look at what data collection strategy would be employed on a sporadic event, we find frequency does NOT work in our favor. Under these circumstances, our detective may be investigating a single homicide and be reliant on the evidence at that scene only. This would mean we must be very diligent about collecting the data from the scene before it is tampered with. When a sporadic event occurs, we must be diligent at that time to collect the data in spite of the massive efforts to get the operation running again.

PRESERVING EVENT DATA

The first step in the PROACT RCA methodology, as is the case in any investigative or analytical process, is to preserve and collect relevant data. Before we discuss the specifics of how to collect various forms of data and when to collect it, let's take a look at the psychological side of why people should assist in collecting data from an event scene.

Let's create a scenario where you are a mechanic in a manufacturing plant. You just completed a 10-day shutdown of the facility to perform scheduled maintenance. Everyone knows at this facility that when the plant manager says the shutdown will last 10 days and no more, you do not want to be the one responsible for extending it past 10 days. A situation arises in the ninth day of the shutdown where, during an internal preventive maintenance inspection, you find that a part has failed and must be replaced. In good faith you request the part from the storeroom. The storeroom personnel inform you that the particular part is out of stock and it will take 4 weeks to expedite the order from the vendor. Knowing this is the ninth day of the 10-day shutdown, you decide to make a "Band-Aid" repair because you do not want to be the person to extend the shutdown. You rationalize that the "Band-Aid" will hold for the 4-week duration as you have gotten away with this in the past. So you install a *not* "like-for-like" part in preparation for the startup of the process.

Within 24 hours of startup the process fails catastrophically and all indications lead to the area where the "Band-Aid" fix was installed. A formal RCA team is amassed and you are assigned to collect some parts data from the scene immediately. Given the witch-hunting culture that you know exists, why should you uncover data/evidence that will incriminate you? While this is a hypothetical scenario, it could very well represent many situations in any industry. "What is the incentive to collect event data in hopes of uncovering the truth"? After all, this is a time-consuming task. It will lead to people who used poor judgment, and therefore management could witch-hunt them and apply certain disciplinary actions.

These are all very valid concerns. We have seen the good, the bad, and the ugly created by these concerns. The fact of the matter is if we wish to uncover the truth, the real root causes, we cannot do so without the necessary data. Think about any investigative or analytical profession—the first step is always to design data collection strategies to obtain the data. Is a detective expected to solve a crime without any evidence or leads? Is an NTSB investigator expected to determine the reasons for an airline accident without any evidence from the scene? Do doctors make diagnoses without any information as to how the patient presented? If these professionals see the necessity of gathering data and information to draw conclusions, then in industry we certainly must recognize the correlation to RCA.

Based on our experience, we have seen a general resistance to data collection for RCA purposes. We can draw two general conclusions from our experience (Figure 8.2):

1. People are resistant to collecting event data because they do not appreciate the value of the data to an analysis or analyst.
2. People are resistant to collecting data because of the paradigms that exist with regard to witch-hunting and managerial expectations.

The first conclusion is the minor of the two. Oftentimes production in any facility is the ruling body. After all, we are paid to produce quality products or services whether

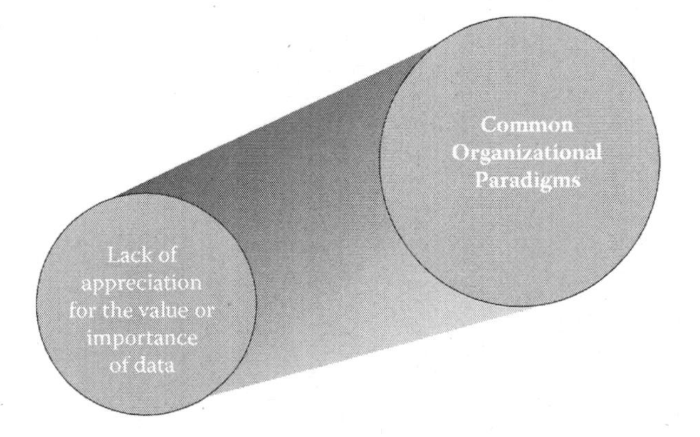

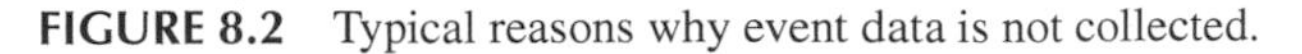
FIGURE 8.2 Typical reasons why event data is not collected.

that product is oil, steel, package delivery, or quality patient care. When this mentality is dominant, it forces us to react with certain behaviors. If production is paramount, then whenever an event occurs, we must clean it up and get production started as quickly as possible. The focus is not on why the event occurred; rather, it is on the fact that it did occur and we must get back to our status quo as quickly as possible.

This paradigm can be overcome merely with awareness and education. Management must first commit to supporting RCA both verbally and on paper. We discussed earlier in the management support chapter that demonstrated actions are seen as "walking the talk," and one of those actions was issuing an RCA policy and/or procedure. This requires data to be collected instead of making it an option. Second, it is not just enough to support the data collection, but we must link with the individuals who must physically collect the data. They must clearly understand "WHY" they should collect the data and "HOW" to do it properly.

We should link with people's value systems and show them the purpose of data collection. If you are an operator in a steel mill and the first one to an event scene, you should understand what is important versus unimportant information to an RCA. For instance, you can view a broken shaft as an item to clean up or as an integral piece of information for a metallurgist. If you understand how important the data you collect is to an analysis, you will see and appreciate why it should be collected. If you do not understand or appreciate its value, then the task is seen as a burden to your already full plate. Providing everyone with basic training in proper data collection procedures can prove invaluable to any organization.

We have seen the potential consequences of poor data collection efforts in some recent high-profile court cases. Allegations are made as to the sloppy handling of evidence in lab work, improper testing procedures, improper labeling, and contaminated samples. Issues of these types can lose *your case* as well.

Providing support and training overcomes one hurdle. But it does not clear the hurdle of perceived witch-hunting by an organization. Many people will choose not to collect data for fear that they may be targeted based on the conclusion drawn from the data. This is a very prominent cultural issue that must be addressed in order to progress with RCA. We cannot determine "Root" causes if a witch-hunting culture is prevalent.

THE ERROR-CHANGE PHENOMENON

Our experience indicates that there are an average number of errors that must queue up in a particular pattern for a catastrophic event to occur. The Error Chain Concept* "… describes human error accidents as the result of a sequence of events that culminate in mishaps. There is seldom an overpowering cause, but rather a number of contributing factors of errors, hence the term *error chain*. Breaking any one link in the chain might potentially break the entire error chain and prevent a mishap." This research comes from the aviation industry and is based on the investigation of more than 30 accidents or incidents. This has been our experience as well in investigating industrial failures.

* Flight Safety International, Crew Resource Management Workshop, September 1993.

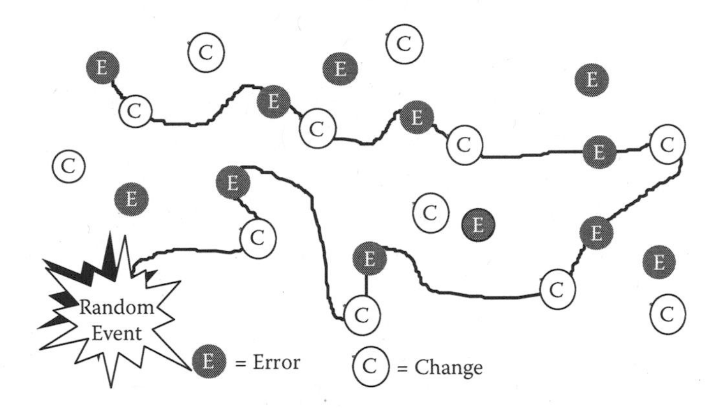

FIGURE 8.3 The Error-Change Phenomenon.

Flight Safety International states that the fewest links discovered in any one accident was four, the average being seven.* Our experience in industrial applications shows the average number of errors that must queue up to be between 10 and 14. To us, this is the core to understanding what an analyst needs in order to understand why undesirable events occur.

We like referring to it as error-change relationships. First, we must define some terms in order to communicate more effectively. We will use James Reason's (*Human Error*, 1990)† definition of human error for our RCA purposes. James Reason defines human error as "… a generic term to encompass all those occasions in which a planned sequence of mental and physical activities fails to achieve its intended outcome, and when these failures cannot be attributed to some chance agency." This means we intended to have a satisfactory outcome and it did not occur. We, in some manner, either (1) deviated from our intended path or (2) the intended path was incorrect.

The *change*, as a result of an error in our environment, is something that is perceptible to the human senses. An example might be that we commit an error by misaligning a shaft. The change will be that an excessive vibration occurs as a result. A nurse administering the wrong medication to a patient is the human error. The adverse reaction is the perceptible change. These series of human errors and associated changes are occurring around us every day. When they queue up in a particular pattern that is when catastrophic occurrences happen (Figure 8.3).

James Reason coined the term Swiss Cheese Model‡ to depict this scenario graphically and this term has caught on in many industries (Figure 8.4).

Knowing this information, we would like to make two points:

1. We as human beings have the ability through our senses to be more aware of our environments. If we sharpen our senses, we can detect these changes and take action to prevent the error chain from running its course. Many

* Flight Safety International, Crew Resource Management Workshop, September 1993.
† Reason, James. 1990. *Human Error*. New York: Cambridge University Press.
‡ Reason, James. 1990–1991. *Human Error*. Victoria: Cambridge University Press.

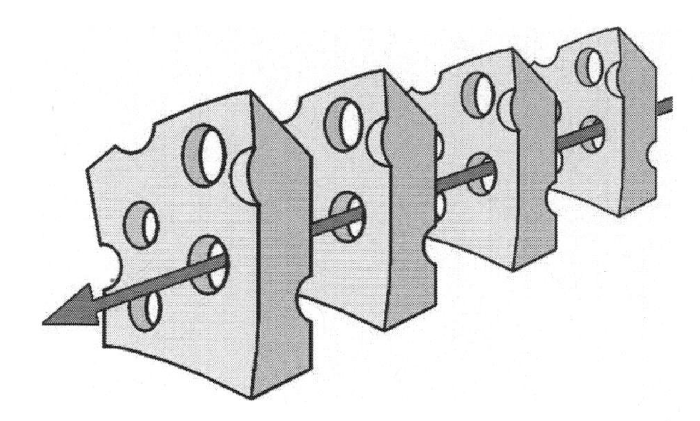

FIGURE 8.4 Reason's Swiss Cheese Model.

of our organizational systems are put in place to recognize these changes. For example, the predictive maintenance group's sole purpose is to utilize testing equipment to identify changes within the process and equipment. If changes are not within acceptable limits, actions are taken to make them within acceptable limits.

2. By witch-hunting the last person associated with an event, we give up the right to the information that person possesses on the other errors that led up to the event. If we discipline a person associated with the event because our culture requires a "head to roll," then that person (or anyone around the person) will likely not be honest about why he or she made the decisions that resulted in errors.

In a later chapter called "Analyzing the Data" (Chapter 10), we will explore what we call a Logic Tree. This is a graphical representation of an error-change chain based on this research. We discuss this research at this point because it is necessary to understand that an investigation or analysis cannot be performed without data. We have enough experience in the field application of RCA to make a general statement that the physical activity of obtaining such data can have many organizational barriers. Once these barriers are recognized and overcome, the task of actually preserving and collecting the data can happen.

THE 5P'S CONCEPT

*Pr*eserving Failure Data is the *PR* in PROACT. In a typical high-profile RCA, an immense amount of data is typically collected and then must be organized and managed. As we go through this discussion we will relate how to manage this process manually versus with software. We will discuss automating your RCA using software technologies in Chapter 13.

Consider this scenario: a major upset just occurred in your facility. You are charged to collect the necessary data for an investigation. What is the necessary information to collect for an investigation or analysis? We use a 5P's approach, where the P's stand for the following:

1. **P**arts
2. **P**osition
3. **P**eople
4. **P**aper
5. **P**aradigms

Virtually anything that needs to be collected from an event scene can be stored under one of these categories. Many items will have shades of gray and fit under two or more categories, but the important thing is to capture the information and slot it under one category. This categorization process will help document and manage the data for the analysis.

Let's use the parallel of the police detective again. What do we see detectives and police officers routinely do at a crime scene? We see the police rope off the area preserving the positional information. We see detectives interviewing people who may be eyewitnesses. We see forensic teams "bagging and tagging" evidence or parts. We see a hunt begin for information or a paper trail of a suspect that may involve past arrests, insurance information, financial situation, etc. And lastly, as a result of the interviews with the observers, we draw tentative conclusions about the situation such as "… he was always at home during the day and away at night. We would see children constantly visiting for five minutes at a time. We think he is a drug dealer." These are the paradigms that people have about situations that are important, because if they believe these paradigms, then they are basing their decisions on them. This can be dangerous.

PARTS

Parts will generally mean something physical or tangible. The potential list is endless, depending on the facility where the RCA is conducted. For a rough sampling of what is meant by parts, please review the following lists:

Continuous process industries (oil, steel, aluminum, paper, chemicals, etc.)

- Bearings
- Seals
- Couplings
- Impellers
- Bolts
- Flanges
- Grease samples
- Product samples
- Water samples
- Tools
- Testing equipment
- Instrumentation
- Tanks
- Compressors
- Motors

Discrete product industries (automobiles, package delivery, bottling lines, etc.)
- Product samples
- Conveyor rollers
- Pumps
- Motors
- Instrumentation
- Processing equipment

Healthcare (hospitals, nursing homes, outpatient care centers, long-term care facilities, etc.)
- Medical diagnostic equipment
- Surgical tools
- Gauze
- Fluid samples
- Blood samples
- Biopsies
- Medications
- Syringes/needles
- Testing equipment
- IV pumps
- Patient beds/rails

This is just a sampling to give you a feel for the type of information that may be considered under the parts category.

Position

Positional data is the least understood and what we consider to be the most important. Positional data comes in the form of two different dimensions, one being physical space and the other being point in time. Positions in terms of space are vitally important to an analysis because of the facts that can be deduced.

When the space shuttle Challenger exploded on January 28, 1986, it was approximately 5 miles in the air. Films from the ground provided millisecond-by-millisecond footage of the parts that were being dispersed from the initial cloud. From this positional information, trajectory information was calculated and search and recovery groups were assigned to approximate the locations of where vital parts were located. Approximately 93,000 square miles of ocean were involved in the search and recovery of shuttle evidence in the government investigation.* While this is an extreme case, it shows how position information is used to determine, among other things, force.

While on the subject of the shuttle Challenger, other positional information that should be considered is "Why was it the right Solid Rocket Booster (SRB) and not the left?" "Why was it the aft (lower) field joint attachment versus the upper field joint attachment?" "Why was the leak at the O-ring on the inside diameter of the

* Challenger: Disaster and Investigation. Cananta Communications Corp., 1987

TABLE 8.1
Space Shuttle Columbia Debris Damage Events

Mission	Date	Comments
STS-1	04/12/81	Lots of debris damage. 300 tiles replaced.
STS-7	06/18/83	First known left bipod ramp foam shedding event.
STS-27R	12/02/98	Debris knocks off tile, structural damage and near burn through results.
STS-32R	01/09/90	Second known bipod event.
STS-35	12/02/90	First time NASA calls foam debris "safety of flight issue," and "re-use or turn-around time issue."
STS-42	01/22/92	First mission after the next mission (STS-45) launched without debris in-flight anomaly closure/resolution.
STS-45	03/24/03	Damage to wing RCC Panel 10-right. Unexplained anomaly, "most likely orbital debris."
STS-50	06/25/92	Third known bipod ramp foam event. Hazard Report 37: Accepted Risk.
STS-52	10/22/92	Undetected bipod ramp foam loss (fourth bipod event).
STS-56	04/08/93	Acreage tile damage (large). Called within "experience base."
STS-62	10/04/94	Undetected bipod ramp foam loss (fifth bipod event).
STS-87	11/19/97	Damage to Orbiter Thermal Protection System spurs NASA to begin 9 flight tests to resolve foam shedding. Foam fix ineffective. In-flight anomaly eventually closed after STS-101 as "accepted risk."
STS-112	10/07/02	Sixth known left bipod ramp foam loss. First time major debris event not assigned an in-flight anomaly. External tank was assigned an action. Not closed out until after STS-113 and STS-107.
STS-107	01/16/03	Columbia Launch. Seventh known left bipod ramp foam loss event.

SRB versus the outside diameter?" These are questions regarding positional information that had to be answered.

Now let's take a look at positions in time and their relative importance. Monitoring *positions in time* in which undesirable outcomes occur can provide information for correlation analysis. By recording historical occurrences we can plot trends that identify the presence of certain variables when these occurrences happen. Let's take a look at the shuttle Challenger again. Most of us remember the incident and the conclusion reported to the public: an O-ring failure resulting in a leak of solid rocket fuel. If we look at the positional information from the standpoint of time, we would learn that the O-rings had evidence of secondary O-ring erosion on 15 of the previous 25 shuttle launches.* When the SRBs are released they are parachuted into the ocean, retrieved, and analyzed for damage. The correlation of these past launches that incurred secondary O-ring erosion showed that low temperatures were a common variable. The *positions in time* information aided in this correlation.

Ironically, in the shuttle Columbia breakup on January 16, 2003, there were seven occurrences of bipod ramp foam events since the first mission STS-1. Table 8.1 identifies which missions incurred which types of damage.

* Lewis, Richard S. 1988. *Challenger: The Final Voyage*. New York: Columbia University Press.

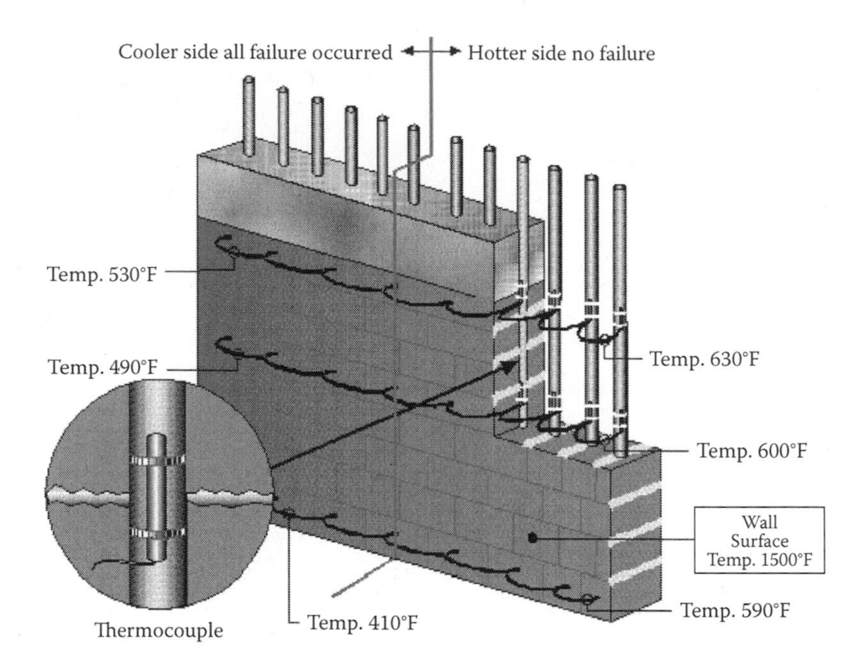

FIGURE 8.5 Mapping example of sulfur burner boiler.

The long and short of it is that the loss of foam tiles from the main fuel tanks and their subsequent impact on the shuttle vehicle were not a new phenomenon—just like the O-ring erosion occurrences. Collecting the positions in time of these occurrences and mapping them out on a time line prove these correlations.

Moving into more familiar environments, we can review some general or common positional information to be collected at almost any organization:

- Physical position of parts at scene of incident
- Point in time of current and past occurrences
- Position of instrument readings
- Position of personnel at time of occurrence(s)
- Position of occurrence in relation to overall facility
- Environmental information related to position of occurrence such as temperature, humidity, wind velocity, etc.

We are not looking to recruit artists for these maps or sketches (Figure 8.5). We are simply seeking to ensure that everyone sees the situation the same way based on the facts at hand. Again, this is just a sampling to get individuals in the right frame of mind of what we mean by positional information.

People

The "People" category is the more easily defined "P". This is simply who we need to talk to initially in order to obtain information about an event. The people we must talk to first should typically be the physical observers or witnesses to the event. Efforts to obtain such interviews should be relentless and immediate. We risk the chance of losing direct observation when we interview observers days after an event occurs. We will ultimately lose some degree of short-term memory and also risk the observers having talked to others about their opinion of what happened. Once observers discuss such an event with another outsider, they will tend to reshape their direct observation with the new perspectives.

We have always identified the goal of an interview with an observer to be that we must be able to see through their eyes what they saw at the scene. The description must be vivid, and it is up to the interviewer to obtain such clarity through the questioning process.

Interviewing skills are necessary in such analytical work. People must feel comfortable around an interviewer and not intimidated. A poor interviewing style can ruin an interview and subsequently an analysis or investigation. A good interviewer will understand the importance and value of body language. Experts estimate approximately 55% to 60% of all communication between people is through body language. Approximately 30% of communication is through the tonal voice and 10% to 15% is through the spoken word.* This is very important when interviewing because it emphasizes the need to interview in person rather than over the telephone. If you look at the legal profession, lawyers are professionals at reading the body language of their clients, their opposition, and the witnesses. Body language clues will direct their next move. This should be the same for interviews associated with an undesirable outcome. The body language will tell interviewers when they are getting close to information they desire, and this will direct the line and tone of subsequent questioning.

Consider another profession that we might not think of as having a strong relationship with body language—professional poker players. It does not take the novice long to realize that the strength of the cards you are dealt does not determine if you are a winner. Professional poker players play their hands based on their read of the body language of their opponents. They know that there are certain involuntary responses of the body by certain players that indicate they are holding a strong hand or that they are likely bluffing. This further validates the importance and effect of body language when interviewing.

When interviewing during the course of an RCA, it is also important to consider the logistics of the interview. Where is the appropriate place to interview? How many people should we interview at a time? What types of people should be in the room at the same time? How will we record all the information? Preparation and environment are very important factors to consider.

We discuss the interviewing environment and the ideal number of people in an interview in Chapter 9. These same pointers will hold true when interviewing for the actual Root Cause Analysis (RCA) versus the Opportunity Analysis.

We have the most success in interviews when the interviewees are from various departments, and more specifically from different "kingdoms." We define kingdoms as entities that build their castles within facilities and tend not to communicate with

* Lyle, Jane. 1990. *Body Language*. London: The Hamlyn Publishing Group Limited.

each other. Examples can be maintenance versus operations, labor versus management, doctors versus nurses, hourly versus salary, etc. When such groups get together they learn a great deal about the others' perspective and tend to earn a respect for each other's position. This is another added benefit of an RCA—people actually start to meet and communicate with others from different levels and areas.

If an interviewer is fortunate enough to have an associate analyst to assist, the associate analyst can take the notes while the interviewer focuses on the interview. It is not recommended that recording devices be used in routine interviews as they are intimidating and people believe that the information may be used against them at a later date. In some instances where significant legal liabilities may be at play, legal counsel may impose such actions. However, if they do, they are generally doing the interviewing. In the case of most chronic failures or events, such extremes are rare.

Typical people to interview will again be based on the nature of the industry and the event being analyzed. As a sample of potential interviewees, consider the following list:

- Observers
- Maintenance personnel
- Operations personnel
- Management personnel
- Administrative personnel
- Technical personnel
- Purchasing personnel
- Storeroom personnel
- Vendor representatives
- Original equipment manufacturers (OEMs)
- Personnel at other similar sites with similar processes
- Inspection/quality control personnel
- Risk/safety personnel
- Environmental personnel
- Lab personnel
- Outside experts

As stated previously, this is just to give you a feel for the variety of people who may provide information about any given event.

PAPER

Paper data is probably the most *understood* form of data. Being in an information age where we have instant access to data through our communications systems, we tend to be able to amass a great deal of paper data. However, we must make sure that we are not collecting paper data for the sake of developing a big file. Some companies seem to feel they are getting paid based on the width of the file folder. We must make sure the data we are collecting is relevant to the analysis at hand.

Keep in mind our detective scenarios discussed earlier and the fact that they are always preparing a solid case for court. Paper data is one of the most effective and expected categories of evidence in court. Solid, organized documentation is the key to a winning strategy.

Typical paper data examples are as follows:

- Chemistry lab reports
- Metallurgical lab reports
- Specifications
- Procedures
- Policies
- Financial reports
- Training records
- Purchasing requisitions/authorizations
- Nondestructive testing results
- Quality control reports
- Employee file information
- Maintenance histories
- Production histories
- Medical histories/patient records
- Safety records information
- Internal memos/e-mails
- Sales contact information
- Process and instrumentation drawings
- Past RCA reports
- Labeling of equipment/products
- Distributive Control System (DCS) strips
- Statistical Process Control/Statistical Quality Control Information (SPC/SQC)

In Chapter 13, "Automating Root Cause Analysis," we will discuss how to keep all this information organized and properly documented in an efficient and effective manner.

Paradigms

Paradigms have been discussed throughout this text as a necessary foundation of understanding how our thought processes affect our problem-solving abilities. But exactly what are paradigms? We will base the definition we use in RCA on futurist Joel Barker's definition, as follows:

> A paradigm is a set of rules and regulations that: 1) Defines boundaries; and 2) tells you what to do to be successful within those boundaries. (Success is measured by the problems you solve using these rules and regulations.)*

This is basically how groups of individuals view the world and react and respond to situations arising around them. This inherently affects how we approach solving problems and will ultimately be responsible for our success or failure in the RCA effort.

---

* Barker, Joel. 1989. *Discovering the Future: The Business of Paradigms*. Elmo, MN: ILI Press.

Paradigms are a by-product of interviews carried out in this process, which were discussed earlier in this chapter. Paradigms are recognizable because repetitive themes are expressed in these interviews from various individuals. How an individual sees the world is a mindset. When a certain population shares the same mindset, it becomes a paradigm. Paradigms are important because even if they are false, they represent the beliefs in which we base our decision making. Therefore, true paradigms represent reality to the people that possess them.

Following is a list of common paradigms we see in our travels. We are not making a judgment as to whether or not they are true, but rather that they affect judgment in decision making.

- We do not have time to perform RCA.
- We say safety is number one, but when it comes down to brass tacks on the floor, cost is really number one.
- This is impossible to solve.
- We have tried to solve this for 20 years.
- It's old equipment; it's supposed to fail.
- We know because we have been here for 25 years.
- This is another program-of-the-month.
- We do not need data to support RCA because we know the answer.
- This is another way for management to "witch-hunt."
- Failure happens; the best we can do is sharpen our response.
- RCA will eliminate maintenance jobs.
- It is a career-limiting choice to contradict the doctor (a nurse's perspective).
- We fully trust the hospitals to be responsible for our care.
- Hospitals are safe havens for the sick.
- What we get is what we order; there is no need to check.
- RCA is RCA; it is all the same.
- We don't need RCA; we know the answer.
- If the failure is compensated for in the budget, it is not really a failure anymore.
- RCA is someone else's job, not mine.

Many of these statements may sound familiar. But think about how each statement could affect problem-solving abilities. Consider the following if-then statements:

- If we see RCA as another burden (and not a tool) on our plate, then we will not give it a high priority.
- If we believe that management values profit more than safety, then we may rationalize at some time that bending the safety rules is really what our management wants us to do.
- If we believe that something is "impossible" to solve, then we will not solve it.
- If we believe that we have not been able to solve the problem in the past, then no one will be able to solve it.
- If we believe that equipment will fail because it is old, then we will be better prepared to replace it.

- If we believe RCA is the program-of-the-month, then we will wait it out until the fad goes away.
- If we do not believe data collection is important, then we will rely on word of mouth and allow ignorance and assumption to penetrate an RCA as fact.
- If we believe that RCA is a witch-hunting tool, then we will not participate.
- If we believe failure is inevitable, then the best we can do is become a better responder.
- If we believe that RCA will eventually eliminate our jobs, then we will not let it succeed.
- If a nurse believes that it is career limiting to contradict a doctor's order, then someone will likely die as a result of the silence.
- If we believe that the hospital is in total control of our care, then we will not question things that seem wrong.
- If we believe that hospitals are safe havens for the sick, then we are stating that we are not responsible for our own safety.
- If we believe that what we get is what we order, then we will not ever inspect when we receive an order and just trust the vendor.
- If we believe that all RCA is the same, then techniques like the 5 Why's will be considered as comprehensive and thorough as PROACT.
- If we believe we know all of the answers, then RCA will not be valued.
- If we believe that unexpected failure is covered for in the budget, then we will not attempt to resolve those unexpected failures.
- If we believe that RCA is someone else's job, then we are indicating that our safety is the responsibility of others and not ourselves.

The purpose of these "if-then" statements is to show the effect that paradigms have on human decision making. When human errors in decision making occur, it is the triggering mechanism for a series of other subsequent errors until the undesirable event surfaces and is recognized.

We have discussed in detail the error-change phenomenon and the 5P's. Now we must discuss how we get all of this information. When an RCA team has been commissioned, a group of data collectors must be assembled to brainstorm what data will be necessary to start the analysis. This first team session is just that, a brainstorming session of data needs. This is not a session to analyze anything. The group must be focused on data needs and not be distracted by the premature search for solutions. The goal of this first session should not be to collect 100% of the data needed. Ideally, our data collection attempts should result in capturing about 60% to 70% of the necessary data. All of the obvious surface data should be collected first and also the most fragile data. Table 8.2 describes the normal fragility of data at a typical event scene. By fragility we mean the prioritization of the 5P's in terms of which is most important to collect first, second, third, and so on. We should be concerned about which data has the greatest likelihood of being tainted the fastest.

TABLE 8.2
Data Fragility Rankings

5P's	Fragility Ranking
Parts	2
Position	1
Paper	3
People	1
Paradigms	4

You will notice that *People* and *Position* are tied for first. This is not an accident. As we discussed earlier in this chapter, the need to interview observers is immediate in order to obtain direct observation. *Positional* information is equally important because it is the most likely to be disturbed the quickest. Therefore, attempts to get such data should be performed immediately. *Parts* are second because if there is not a plan to obtain them, they will typically end up in the trash can. *Paper* data is generally static with the exception of process or on-line production data (DCS, SPC/SQC). Such technologies allow for automatic averaging of data to the point that if the information is not retrieved within a certain time frame, it can be lost forever. *Paradigms* are last because we wish we could change them faster, but modifying behavior and belief systems takes more time.

One preparatory step for analysts should be to always have a data collection kit prepared. Many times such events occur when we least expect it. We do not want to have to run around collecting a camera, plastic bags, etc. If it is all in one place it is much easier to be prepared in a minute's notice. Usually good models are from other emergency response occupations such as doctor's bags, fire departments, police departments, EMTs, etc. They always have most of what they need accessible at any time. Such a bag (in general) may have the following items:

- Caution tape
- Masking tape
- Plastic Ziploc® bags
- Gloves
- Safety glasses
- Ear plugs
- Adhesive labels
- Marking pens
- Digital camera w/spare batteries
- Video camera (if possible)
- Marking paint
- Tweezers
- Pad and pen
- Measuring tape
- Sample vials
- Wire tags to ID equipment

This is, of course, a partial listing and depending on the organization and nature of work, other items would be added or deleted from the list.

The form in Figure 8.6 is a typical data collection form used for manually organizing data collection strategies for an RCA team.

1. Data Type/Category—Which of the 5P's this form is directed at is listed. Each "P" should have its own form.
2. Person Responsible—The person responsible for making sure the data is collected by the assigned date.

5P's Data Collection Form

Analysis Name: ______________________________

Data Type: People, Parts, Position, Paper, Paradigms (circle one)

Champion: ______________________________

(Person who ensures all data assigned below is collected by due date)

#	Data to Be Collected	How Data Will Be Obtained (Data Collection Strategy)	Person Responsible	Date to Be Collected By

FIGURE 8.6 Sample 5P's Data Collection Form.

3. Data to Collect—During the 5P's brainstorming session, list all data necessary to collect for each "P".
4. Data Collection Strategy—This space is for actually listing the plan of how to obtain the previously identified *data to collect*.
5. Date to Be Collected By—Date by which the data is to be collected and ready to be reported to team.

Figure 8.7 shows a sample completed Data Collection Form.

5P's Data Collection Form

Analysis Name: Recurring Failure of Pump 235
Data Type: People, Parts, Position, (Paper,) Paradigms (circle one)
Champion: John Smith
(Person who ensures all data assigned below is collected by due date)

#	Data to Be Collected	How Data Will Be Obtained (Data Collection Strategy)	Person Responsible	Date to Be Collected By
1	Shift Logs	Have shift foreman collect the shift logs when pump 235 fails and deliver it to John Smith within 1 day.	Ken Latino	11/30/05

FIGURE 8.7 Completed Data Collection Form.

9 Ordering the Analysis Team

When a sporadic event typically occurs in an organization, an immediate effort is organized to form a task team to investigate "WHY" such an undesirable event occurred. What is the typical make-up of such a task team? We see the natural tendency of management to assign the "cream-of-the-crop" experts to both lead and participate on such a team. While well intended, there are some potential disadvantages to this thought process.

Let's paint a scenario in a manufacturing setting (even though it could happen anywhere). A sulfur burner boiler fails due to tube ruptures. The event considerably impacts production capabilities when it occurs. Maintenance histories confirm that such an occurrence is chronic as it has happened at least once a year for the past 10 years. Therefore, Mean-Time-Between-Failure (MTBF) is approximately one per year. This event is a high priority on the mind of the plant manager, as it is impacting his facility's ability to meet corporate production goals and customer demand in a reliable manner. He is anxious for the problem to go away. He makes the logical deduction that if he has tubes rupturing in this boiler, then it must be a metals issue. Based on this premise he naturally would want to have his best people on the team. He assigns his top metallurgist as the team leader because he has been with the company the longest and has the most experience in the materials lab. On the team he will provide the metallurgist the resources of his immediate staff to dedicate the time to solve the problem. Does this sound familiar? The logic appears sound. Why wouldn't this strategy work?

Let's review what typically happens next. We have a team of say five metallurgists. They are brainstorming all the reasons these tubes could be bursting. At the end of their session(s) they conclude that more exotic metals are required and that the tube materials should be changed in order to be able to endure the harsh atmosphere in which they operate. Problem solved! However, this is the same scenario that went on for the past 10 years and they kept replacing the tubes year after year and the tubes still kept rupturing.

Think about what just went on with that team. Remember our earlier discussion about *paradigms* and how people view the world. How do we think the team of metallurgists view the world? They all share the same "box." They have similar educational backgrounds, similar experiences, similar successes, and similar training. That is what they know best: metallurgy. Any time we put five metallurgists on a team we will typically have a metallurgical solution. The same goes for any expertise in any discipline. This is the danger of not having technical diversity on a team and also of letting an expert lead a team on an event in which the team members are the experts. Our greatest intellectual strengths represent liabilities when they lead us to miss something that we might have otherwise noticed—they create *blind spots*.*

The end to the story above is that eventually an engineer of a different discipline was assigned as the leader of the team. The new team had metallurgists as well as mechanical and process engineers. The end result of the thorough RCA was that the tubes that were rupturing were in a specific location of the boiler that was below the dew point for sulfuric acid. Therefore, the tubes were corroding due to the environment. The solution: return to the base metals and move the tubes 18 inches forward (outside of the brick wall) where the temperature was within acceptable limits.

When team leaders are NOT experts, they can ask any question they wish of the team members who should be the experts. However, this luxury is not afforded to experts who lead RCA teams because their team members generally perceive them as all-knowing. Therefore, they cannot ask the seemingly obvious or *stupid* question. While this seems a trivial point, it can, in fact, be a major barrier to success.

NOVICES VERSUS VETERANS

As much as management would like to believe that sending their personnel to RCA would result in them leaving the classroom as experts, this is not to be. Like anything that we become proficient in, it requires practice. We must realize that learning a structure process like PROACT involves changing the thought patterns to which we are accustomed. This does not happen easily or quickly.

Should novices be using a different RCA approach than that used by veterans? No. How well any given approach is used will determine how effective it is. Novices tend to be *skittish* at first and uncomfortable with the change in thinking as a whole. Therefore, they may tend to take shortcuts or overlook some steps to accelerate progress. They may tend to let their aggressive team members intimidate them as team leaders, and this may result in them accepting hearsay as fact.

Novices may choose not to be as disciplined at data collection because of the time it takes to collect the data. They may not be proficient at interviewing people under stressful conditions and therefore not uncover the information they would like. A novice's logic tree will likely have *logic holes* or gaps in logic because the novice lacks experience. However, aren't these all just small signals of inexperience? Don't they happen with anything new that we learn?

Novices will gradually becomes veterans by jumping into it and giving it their best shot. They will recognize that their primary role is to stick to the adherence of the discipline of the RCA approach. This will be in light of the obstacles that they will inevitably face due to the culture of the organization. Novices will make mistakes and then they will be stronger as a result. Novices should recognize that they are novices and not become overconfident in their capabilities in the early stages. Again, overconfidence is one of the leading causes of human error.

---

* Van Hecke, Madeleine L. 2007. *Blind Spots: Why Smart People Do Dumb Things*. New York: Prometheus Books, p. 22.

As mentioned in the introductory chapter, perhaps organizations will develop RCA procedures that will guide this RCA Principal Analyst maturation process. If we had a procedure and development plan, maybe they would require specific training before being included on an RCA team as a team member. Then, after having been a team member on six analyses, they are eligible to be cofacilitators. They would then serve as cofacilitators on another six analyses before they would be eligible to go solo and lead a team.

Novices should start off with analyses that have a reasonably good chance of success. They should not try to conquer world hunger on the first attempt. They should strive to build confidence in their capabilities after each analysis. This growing confidence will make them veteran analysts with a solid foundation in the principles of true RCA.

THE RCA TEAM

To avoid this trap of narrow-minded thinking, let's explore the anatomy of an ideal RCA team. The purpose of a diverse team is to provide synergism where the whole is greater than the sum of its parts. Anyone who has participated in the survival-type teaming games and outings will agree—when different people of different backgrounds come together for a team purpose, their outcomes are better as a team than if they had pursued the problem as individuals.

Teams have long been a part of the quality era and are now commonplace in most organizations. Working in a team can be the most difficult part of our work environment because we will be working with others who may not agree with our views. This is the reason that teams work—people disagree. When people disagree, each side must make a case to the other why its perspective is correct. To support this view, a factual basis must be provided rather than "conventional wisdom." This is where the learning comes in and teams progress. We always use the line that "if a team is moving along in perfect harmony, then changes need to be made in team make-up." We must seek the necessary debate required to make a team progress. While this may seem difficult to deal with, it will ultimately promote the success of the team's charter.

WHAT IS A TEAM?

> A team is a small number of people with complementary skills who are committed to a common purpose, performance goals, and approach for which they hold themselves mutually accountable.*

A team is different than a group. A group can give the appearance of a team; however, the members act individually rather than in unison with others.

---

* Katzenbach, Jon R. and Smith, Douglas K. 1994. *The Wisdom of Teams.* Boston, MA: Harvard Business School Press.

Let's explore the following key elements of an ideal RCA team structure:

1. Team Member Roles and Responsibilities
2. Principal Analyst Characteristics
3. The Challenges of RCA Facilitation
4. Promote Listening Skills
5. Team Codes of Conduct
6. Team Charter
7. Team Critical Success Factors
8. Team Meeting Schedules

Team Member Roles and Responsibilities

Many views about ideal team size exist. The situation that created the team will generally determine how many members are appropriate. However, from an average standpoint for RCA, it has been our experience that between three and five core team members is ideal and beyond 10 is too many. Having too many people on a team can force the goals to be prolonged due to the dragging on of too many opinions.

Who are the core members of an RCA team? They are as follows:

a. The Principal Analyst (PA)
b. The Associate Analyst
c. The Experts
d. Vendors
e. Critics

The Principal Analyst (PA)

Each RCA team needs a leader. This is the person who will ultimately be held accountable by management for results. They are the people who will drive success and accept nothing less. It is their desire that will either make or break the team. The PA should also be a facilitator, not a participator. This is a very important distinction because the technical experts who lead teams tend to participate instead of facilitate. The PA, as a facilitator only, recognizes that the answers are within the team members, and it is the PA's job to extract those answers in a disciplined manner by adhering to the PROACT methodology.

This person is responsible for the administration of the team efforts, the facilitation of the team members according to the PROACT philosophy, and the communication of goals and objectives to management oversight personnel.

The Associate Analyst

This position is often seen as optional; however, if the resources are available to fill it, it is of great value. The Associate Analyst is basically the *legman* for the PA. This person will execute many of the administrative responsibilities of the PA such as inputting data, issuing meeting minutes, arranging for conference facilities, arranging for audio/visual equipment, obtaining paper data such as records, etc. This person relieves much of the administrative burdens from the PA, allowing the PA more time to focus on team progress.

The Experts

The experts are basically the core make-up of the team. These are the individuals that the PA will facilitate. They are the *nuts-and-bolts* experts on the issue being analyzed. These individuals will be chosen based on their backgrounds in relation to the issue being analyzed. For instance, if we are analyzing an equipment breakdown in a plant we may choose to have operations, maintenance, and engineering personnel represented on the team. If we are exploring an undesirable outcome in a hospital setting we may wish to have doctors, nurses, lab personnel, and quality/risk management personnel on the team. In order to develop accurate hypotheses, experts are absolutely necessary on the team. Experts will aid the team in generating hypotheses and also verifying them in the field.

Vendors

Vendors are an excellent source of information about their products. However, in our opinion, they should not lead an analysis when their products are involved in an event. Under such circumstances, we want the conclusions drawn by the team to be unbiased so that they have credibility. It is often very difficult for a vendor to be unbiased about how its product performed in the field. For this reason, we suggest that vendors participate on the team, but not lead the team.

Vendors are great sources of information for generating hypotheses about how their products could not perform to expectations. However, they should not be permitted to prove or disprove their own hypotheses. We often see situations where the vendor will blame the way in which the product was handled or maintained as the cause of its nonperformance. It always seems to be something the customer did rather than a flaw in the product itself. We are not saying that the customer is always right, but from an unbiased standpoint, we must explore both possibilities: that the product has a problem as well as that the customer could have done something wrong to the product. Remember, facts lead such analyses, not assumptions!

Critics

We have never come across a situation in our careers where we had difficulty in locating critics. Every critic knows who he or she is in the organization. However, sometimes critics get a bad reputation just because they are curious. Critics are typically people who do not see the world the way that everyone else does. They are really the "devil's advocates." They will force the team to see the other side of the tracks and find holes in logic by asking persistent questions. They are often viewed as uncooperative and not team players. But they are a necessity to a team.

Critics come in two forms: (1) constructive and (2) destructive. Constructive critics are essential to success and are naturally inquisitive individuals who take nothing (or very little) at face value. Destructive critics stifle team progress and are more interested in overtime and donuts versus successfully accomplishing the team charter.

Principal Analyst Characteristics

The PA typically has a hard row to hoe. If RCA is not part of the culture, PAs are going against the grain of the organization. This can be very difficult to deal with if PAs are people who have difficulty in dealing with barriers to success. Over the years we have noted the personality traits that make certain PAs stars whereas others have not progressed. Following are the key traits that our most successful analysts portray (many of them led the analyses listed in the case histories of this text):

Unbiased

While we discussed this issue earlier, this is a key trait to the success of any RCA. The leader of an RCA should have nothing to lose and nothing to gain by the outcome of the RCA. This ensures that the outcomes are untainted and credible.

Persistent

Individuals who are successful as PAs are those who do not give up in the face of adversity. They do not retreat at the first sign of resistance. When they see roadblocks, they immediately plan to go through them or around them. "No" is not an acceptable answer. "Impossible" is not in their vocabulary. They are painstakingly persistent and tenacious.

Organized

PAs are required to maintain the organizational process of the RCA. They are responsible for organizing all the information being collected by the team members and putting it into an acceptable format for documentation and presentation. Such skills are extremely helpful in RCA. As mentioned earlier, if an associate analyst is available, he or she will play a major role in assisting the PA in this area.

Diplomatic

Undoubtedly PAs will encounter situations where upper-level management or lower-level individuals will not support or cooperate in the RCA effort. Whether it is maintenance not cooperating with operations, unions boycotting teaming, administration not willing to provide information, or doctors not willing to participate on teams, political situations *will* arise. A great PA will know how to handle such situations with diplomacy, tactfulness, and candidness. The overall objective in all these situations is to get what we want. We work backward from that point in determining the means to attain the end.

The Challenges of RCA Facilitation

Those who have facilitated any type of team can surely appreciate the need to possess the characteristics described above. You can also appreciate the experience that such tasks provide in dealing with human beings. Next, we explore common challenges faced when facilitating a typical RCA team:

Bypassing the RCA Discipline and Going Straight to a Solution

As we all have experienced in our daily routines, the pressure of the daily production can overshadow our intentions of doing things right and stepping back and looking at the big picture. This phenomenon becomes apparent when we organize an RCA team that is well versed in how to repair things quickly to get production up and running again. Such teams will be inclined to pressure the RCA facilitator to hurry up and implement their solution(s). We must keep in mind that experts shine in the details or the "micro" side of the analysis. Experts tend to have difficulty when instructed to think more broadly in more macro terms. This change in thinking will be addressed in detail in Chapter 10, "Analyzing the Data."

Floundering of Team Members

One of the more predominant problems with most RCA attempts is lack of discipline and direction of method. This results in the team becoming frustrated because it appears that the team is going around in circles and getting nothing done. Also, if team members are employed who have not been educated in the RCA methodology, they can see no "light at the end of the tunnel." Such team members tend to get bored quickly and lose interest. At this point, like a fish on the dock, the team flounders.

Acceptance of Opinions as Facts

This often occurs when using methodologies that promote solutions before proving that hypotheses are factual. We have all experienced being so pressured to get back to normal or the status quo that we tend to accept people's opinions as facts so that we can come to consensus quickly and try to implement solutions. Often this haste results in spending money that does not solve the problem and is akin to the "trial-and-error" approach. Techniques such as the 5 Why's, Fishbone, and Brainstorming tend to rely more on hearsay than on using evidence to support hypotheses.

Dominating Team Members

This is generally true of most teams that are organized under any circumstances—there is usually one strong-willed person who tends to impose his or her personality on the rest of the team members. This can result in the other team members being intimidated and not participating (or at least not as openly as they otherwise would), but more likely it pressures the team to accept opinion as fact.

Reluctant Team Members

We have all participated on teams where some members were much more introverted than others. It is not that they do not have the experience or talent to contribute, but their personality is simply not an outgoing one. Sometimes people are reluctant to participate because they feel that authority is in the room and they do not want to appear as though they are not asking the right questions, so they say nothing at all and "do not rock the boat." Other times reluctant team members are that way because they know the truth and are worried about exposing it because someone may get in trouble.

Going Off on Tangents

Again, this can (and does) happen on any team. This is a function of team dynamics that happens when humans work together. The RCA facilitator is charged with sticking to the discipline of the RCA method. This includes keeping the team on track and not letting the focus drift.

Arguing among Team Members

Nothing can be more detrimental to a team than its members engaging in destructive arguments due to closed-mindedness. There is a clear difference between argument and debate. Arguments tend to get polarized and each side takes a firm stance and will not budge. The goal of an argument in these cases is for one side to agree with the other totally, not to come to consensus. Debate promotes consensus, which requires a willingness to meet in the middle.

Promote Listening Skills

Obviously many of the team dynamics issues that we are discussing are not just pertinent to RCA, but to any team. While the concept of listening seems simplistic, most of us are not adept at its use.

Many of us often state that we are not good at remembering names. If we look back at a major cause of this, it is because we never actually listen to people when they introduce themselves to us for the first time. Most of the time when someone introduces him- or herself to us, we are more preoccupied with preparing our response than actually listening to what the person is saying.

Next time you meet someone, concentrate on actually listening to the person's introduction and take an imaginary *snapshot* of the person's face with your eyes. You will be amazed at how that impression will log into your long-term memory and pop up the next time you see that person.

The following are listening techniques that may be helpful when organizing RCA teams:

One Person Speaks at a Time

This may appear to be common sense and a mere matter of respect, but how often do you see this rule broken? We obviously cannot be listening if there is input from more than one person at a time.

Don't Interrupt

Not to mention that this is rude, but let people finish their point while you listen. You will have plenty of time to formulate an educated response. Sometimes we think that if we make statements the fastest and the loudest we will gain ground. We can watch the *Jerry Springer Show* and know this is not the case.

React to Ideas, Not People

This is a very important point and should not be forgotten. Even if you disagree with other team members, NEVER make it a personal issue. We may disagree about someone's ideas, but that does not mean it is a personal issue between them and us. This is totally unproductive and will cause digression rather than progression if permitted to happen.

Separate Facts from Conventional Wisdom

Just as in the courtroom, in our debates we must separate facts from conventional wisdom. After all, in RCA the entire discipline is based on facts. Conventional wisdom originates from opinions and, if not proven, will result in assumptions treated as fact.

Team Codes of Conduct

Codes of conduct were most popular within the Quality circles and the push for teaming. They vary from company to company, but what they all have in common is the desire to make meetings more efficient and effective. Codes of conduct are merely sets of guidelines by which a team agrees to operate. Such guidelines are designed to enhance the productivity of team meetings. Following are a few examples of common sense codes of conduct:

- All members will be on time for scheduled meetings.
- All meetings will have an agenda that will be followed.
- Everyone's ideas will be heard.
- Only one person speaks at a time.
- "3 Knock" Rule Will Apply—This is where a person politely knocks on the table to provide an audio indicator that the speaker is going off track of the agenda topic being discussed.
- "Holding Area"—This is a place on the easel pad where topics are placed for consideration on the next meeting agenda because it is not an appropriate topic for the meeting at hand.

This is just a sampling to give you an idea as to what team meeting guidelines can be like. Many of our clients that have embraced the Quality philosophy will have such codes of conduct framed and posted in all of their conference rooms. This provides a visual reminder that will encourage people to abide by such guidelines in an effort not to waste people's time.

Team Charter/Mission

The team's charter (or sometimes referred to as a Mission) is a one-paragraph statement delineating why the team was formed in the first place. This statement serves as the focal point for the team. Such a statement should be agreed upon not only by the team, but also by the managers overseeing the team's activities. This will align everyone's expectations as to the team's direction and expected results.

The following is a sample team charter reflecting a team that was organized to analyze a mechanical failure:

> To identify the root causes of the ongoing motor failures, occurring on pump CP-220, which includes identifying deficiencies in, or lack of, organizational systems. Appro-

priate recommendations for root causes will be communicated to management for rapid resolution.

Team Critical Success Factors (CSFs)

Critical Success Factors are guidelines by which we will know that we are successful. I have heard CSFs also referred to as Key Performance Indicators (KPIs) and other nomenclature. Regardless of what we call them, we should set some parameters that will define the success of the RCA team's efforts. This should not be an effort in futility in listing a hundred different items. We recommend that no more than eight should be designated per analysis. Experience supports that typically many are used over and over again on various RCA teams. Following are a few samples of CSFs:

- A disciplined RCA approach will be utilized and adhered to.
- A cross-functional section of site personnel/experts will participate in the analysis.
- All analysis hypotheses will be verified or disproven with factual data.
- Management agrees to fairly evaluate the analysis team's findings and recommendations upon completion of the RCA.
- No one will be disciplined for honest mistakes.
- A measurement process will be used to track the progress of implemented recommendations.

Team Meeting Schedules

We are often asked, "What is the average time or duration of an RCA?" The answer is another question, like "How important is the resolution of the event?" The higher the priority of the event being analyzed, the quicker the analysis process will move. We have seen high priority given to events to the degree that full-time teams are assigned and resources and funds are unlimited to find the causes. These are usually situations where there must be a visual demonstration of commitment on behalf of the company because the nature of the event was picked up by the media and the public wants an answer. These are usually analyses of sporadic events versus chronic. The space shuttle Challenger and Columbia disasters are such examples where the public's desire to know forced an unrelenting commitment on behalf of the government to get to the truth.

Unfortunately, such attention is rarely given to events that do not hurt individuals, do not destroy equipment, and do not require analysis due to regulatory compliance. These are usually indicative of chronic events.

TEAM PROCESS

As we tell our students, we provide the architecture of an RCA methodology. It will not work the same in every organization. The model or framework should be molded to each culture into which it is being forced. In essence, we must all play with the hand we are dealt. We do the best we can with what we have.

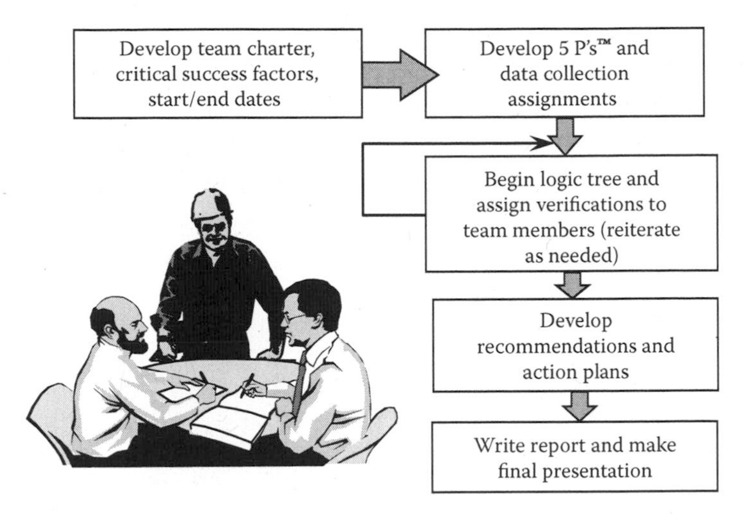

FIGURE 9.1 RCA team process flowchart.

To that end, the process flow involved with such team activities might look like that presented in Figure 9.1.

We can speak ideally about how RCA teams should function, but rarely are there ideal situations in the real world. We have discussed throughout this text the effects of "Re-Engineering" on corporate America and how resources and capital are tight while financial expectations rise. This environment does not make a strong case for organizing teams to analyze why things go wrong.

TEAM APPROACH TO CHRONIC EVENTS

So let's bring this subject to an end with a reality check of how RCA teams will perform under the described conditions. Let's review the analysis of chronic events and how teams will realistically deal with them. Remember that the chronic events are typically viewed as acceptable, part of the budget, and generally do not hurt people or cause massive amounts of damage to equipment. However, they cost the organization the most in losses on an annual basis.

Assume that an Opportunity Analysis (OA) has been performed. The "Significant Few" candidates have been determined (the 80/20 rule). These will likely be chronic versus sporadic events. A team has been formed utilizing the principles outlined in this chapter. Where do we go from here?

The first meeting of an RCA team should be to define the structure of the team and delineate the team's focus. As described in this chapter, the team should first meet to develop its Charter/Mission, Critical Success Factors (CSFs), and the anticipated start and completion dates for the analysis. This session will usually last anywhere from 1 to 2 hours. At the conclusion of this meeting the team should set the next meeting date as soon as possible.

As discussed earlier, because the nature of the events is chronic, we have in our favor frequency of occurrence. From a data collection standpoint, this means opportunity because the event is likely to occur again. Knowing the occurrence is likely to happen again, we can *plan* to collect data about the event. This brings us to the second meeting of the team, whose purpose is to develop a data collection strategy as described in Chapter 8, "Preserving Event Data" through a brainstorming session. This meeting typically will take about 1 to 2 hours and should be scheduled when convenient to the team members' schedules. The result of this meeting will be assignments for members to collect various types of information by a certain date. At the end of this meeting, the next team meeting should be scheduled. The time frame will be dependent on when the information can be realistically collected.

The next meeting will be the first of several involving the delineation of logic utilizing the "Logic Tree" described in Chapter 10. These sessions are reiterative and involve the thinking out of "cause-and-effect" or "error-change" relationships. The first meeting concerning the logic tree development will involve about 2 hours of developing logic paths. It has been our experience that the team should only drive down about 3 to 4 levels on the tree per meeting. This is typically where the necessary data begins to dwindle and hypotheses require more data in order to prove or disprove them. The first tree-building session will incorporate the data collected from the team's brainstorming session on data collection. The entire meeting usually takes about 4 hours. We find that about 2 hours is spent on developing the logic tree and another 2 hours is spent on applying verification information to each hypothesis. At the conclusion of this meeting, a new set of assignments will emerge where verification tests and completion dates will have to be applied to prove or disprove hypotheses. At the conclusion of such logic-tree-building sessions the next meeting date should be set based on the reality of when such verifications can be completed.

Our typical logic tree spans anywhere from 10 to 14 levels of logic. This coincides with the "Error-Change Phenomenon" described in Chapter 8. This means that approximately 3 to 4 logic-tree-building sessions will be required to complete the tree and arrive at the root causes. To recap, this means the team will meet on an as-needed basis three to four times for about 4 hours each in order to complete the logic tree. We are trying to disprove the myth that such RCA teams are taken out of the field full time for weeks on end. We do not want to mislead at this point; we are talking time spent with team members meeting with each other. This is minimal time relative to the time required in the field to actually collect the assigned data and perform the required tests. Proving and disproving hypotheses in the field, by far, is the most time-consuming task in such an analysis. But it is also the most important task if the analysis is to draw accurate conclusions.

By the end of the last logic-tree-building session, all the root causes have been identified and the next meeting date has been set. The next meeting will involve the assigning of team members to write recommendations or countermeasures for each identified root cause. The teams as a whole will review these recommendations; they will then strive for consensus. At the conclusion of this meeting, the final meeting date will be set.

The last team meeting will involve the writing of the report and the development of the final presentation. This meeting may require at least 1 day because we are preparing for our *day in court* and we want to have our solid case ready. Typically, the Principal Analyst will have the chore of writing the report for review by the entire team. The team will work on the development of a professional final presentation. Each team member should take a role in the final presentation to show unity in purpose for the team as a whole. The development of the final report and presentation will be discussed at length in Chapter 11.

As you can see, we have to deal with the reality of our environments. Keep in mind that the above-described process is “an average” for a chronic event. If someone in authority pinpoints any event as a high priority, this process tends to move much faster as support tends to be offered rather than having to fight for it.

In the next chapter, we will move into the details of actually taking the pieces of the puzzle (the data collected) with the ideal team assigned and making sense of a seemingly chaotic situation.

10 Analyzing the Data *Introducing the PROACT Logic Tree*

No matter what methods are out in the marketplace to conduct Root Cause Analysis (RCA), they all use either a categorical or cause-and-effect approach to determining causation. The various RCA methods in the marketplace may vary in presentation, but the legitimate ones are merely different in the way in which they graphically represent the determination of causation. Everyone will have their favorite RCA *tool*, which is fine, as long as they are using it properly and it is producing positive results.

CATEGORICAL VERSUS CAUSE-AND-EFFECT RCA TOOLS

Let's start with exploring the technical aspects of some of the more common RCA tools in the marketplace and contrasting them to each other. We will speak in generalities about these tools, as there is wide variability in how they are applied. Let's explore the following common RCA tools used in the marketplace today:

1. The 5 Why's
2. The Fishbone Diagram
3. The Regulatory Forms
4. The Logic Tree

ANALYTICAL TOOLS REVIEW

The goal of this description is not to teach you how to use these tools properly, but to demonstrate how they can lack breadth and depth of approach and therefore affect the comprehensiveness of the analysis outcomes. Analytical tools are only as good as their users. Used properly and comprehensively, any of these tools can produce good yet variable results. However, experience shows that the attractiveness of these tools is actually their drawback as well. These tools are typically attractive because they are quick to produce a result, require few resources, and are inexpensive. These are the very same reasons they often lack breadth and depth.

Let's start with the 5 Why's. While there are varying forms of this simplistic approach, the most common understanding is that the analyst is to ask him- or herself the question "WHY?" five times and this will uncover *the* root cause. The form of this approach may be as shown in Figure 10.1.

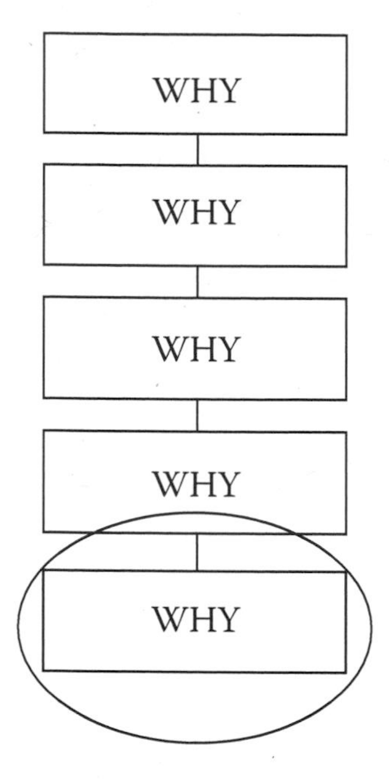

FIGURE 10.1 The 5 Why's analytical tool.

There is a reason we do not hear about NTSB investigators using the 5 Why's approach in the course of their investigations. The main deficiencies with this concept are that failure does not always occur in a linear pattern. Multiple factors combine laterally (parallel) to allow the undesirable outcomes to occur. Also, there is almost never a single root cause, and this is a misleading aspect of this approach as well. People tend to use this tool as individuals and not in a team and rarely back up their assertions with evidence. The original intent of the development of the 5 Why's was for use by individuals working on the Toyota assembly lines. When faced with an undesirable outcome at the line level, individuals were encouraged to think deeper than they normally would to explore possible contributing factors to the outcome. If this did not resolve the issue it would be passed on to a team that would look into the issue using more comprehensive tools.

The fishbone diagram is the second popular analytical Quality tool on the market. This approach gets its name from its form, which is in the shape of a fish (Figure 10.2). The spine of the fish represents the sequence of events leading to the undesirable outcome. The fish bones themselves represent cause categories that should be evaluated as having been potential contributors to the sequence of events. These categories change from user to user. The more popular cause category sets tend to be

- The 6 M's: Management, Man, Method, Machine, Material, Measurement
- The 4 P's: Place, Procedure, People, Policies
- The 4 S's: Surroundings, Suppliers, Systems, Skills

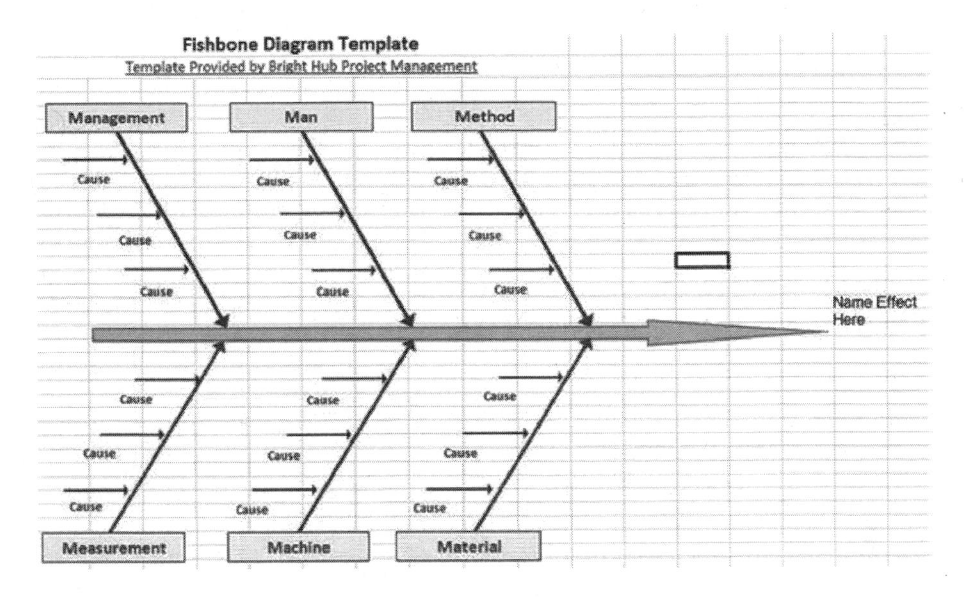

FIGURE 10.2 The Fishbone Diagram sample.

The fishbone diagram is a tool often used in conjunction with brainstorming. Team members decide on the cause category set to be used and continue to ask what factors within the category caused the event to occur. Once these factors are identified and consensus is attained, attention is focused on solutions.

As a brainstorming technique, this tool is less likely to depend on evidence to support hypotheses and more likely to let hearsay fly as fact. This process is also not cause-and-effect based, but category based. The users must pick the category set they wish to use and offer ideas within that category. If the correct categories for the event at hand are not selected, key causes could be overlooked.

The regulatory forms are the last of the popular choices for conducting an RCA. Regulatory forms are often favored, not because of their substance, but because of the perception that if the form is used, there is a greater chance of approval by the agency. There is probably merit to this assumption because by presenting an analysis in another format, even if it is more comprehensive, it may not be viewed as acceptable by the affected agency and it would be looked upon as more work on the agency's part to make it fit their mold (their system).

Many regulatory forms have similar formats. The first portion of the form normally deals with outcomes data. This is information related to the undesirable outcome or the consequences of the event. The second portion of the form deals with determining causation. Most of the time this means that a series of questions (the same questions no matter the incident—one-size-fits-all type of approach) are asked or a list of cause categories are provided and the analyst is expected to brainstorm within those categories. This is very similar to the fishbone approach described earlier.

The third portion of the RCA Regulatory Form is often the Corrective Action portion of the form. After causation is determined, solutions are developed to overcome the identified causes within the categories.

Regulatory forms are a necessary evil when the success of the RCA effort is measured based on compliance. However, when measuring success based on a bottom-line metric, they often fall short. For instance, we may be compliant, but does "being compliant" mean

1. There has been a decrease in the undesirable behaviors that led to the bad outcome?
2. There has been a decrease in affected operations and maintenance expenses?
3. There has been an increase in throughput?
4. There has been an increase in the quality of the product and therefore value to the customer?
5. There has been an increase in client satisfaction?
6. There has been an increase in profitability?

When measuring the success of an RCA, we want to ensure that the metrics that measure our success reflect an improvement to the overall organization's goals. While compliance is certainly going to be a primary goal, we should get a more significant bottom-line benefit from being compliant. Identification of that benefit is a key in the success of our RCA effort.

The PROACT Logic Tree is representative of a tool specifically designed for use within RCA. The logic tree is an expression of cause-and-effect relationships that queued up in a particular sequence to cause an undesirable outcome to occur. These cause-and-effect relationships are validated with hard evidence as opposed to hear-say. The data leads the analysis—not the loudest expert in the room. The strength of the tool is such that it can, and is, used in court to represent solid cases.

This chapter is about the construction of a Logic Tree during the course of an RCA using the PROACT methodology rule sets. We will elaborate on the details of the logic and demonstrate that the use of a comprehensive cause-and-effect tool like the Logic Tree will identify causes that normally would not be picked up in a categorical RCA approach.

Comprehensive cause-and-effect tools allow analysts the opportunity to directly correlate deficient systems to poor decision making to undesirable outcomes using hard evidence. When using categorical approaches where we brainstorm about what we *think* happened within a cause category, usually we cannot directly correlate that cause to a poor decision that led to a bad outcome. When properly using a cause-and-effect tool like the Logic Tree, it will drive us to identify the cause categories that actually played a role in the bad outcome, rather than our having to guess when using a categorical RCA tool.

THE GERMINATION OF A FAILURE

Before getting into a detailed discussion about how to graphically express the sequence of events leading to an undesirable outcome, let's first briefly discuss the origin of a failure and the path it takes to the point that we have to do something about it.

Figure 10.3 expresses the origin of failure as coming from deficient organizational systems. Such systems include policies, procedures, training systems, purchasing

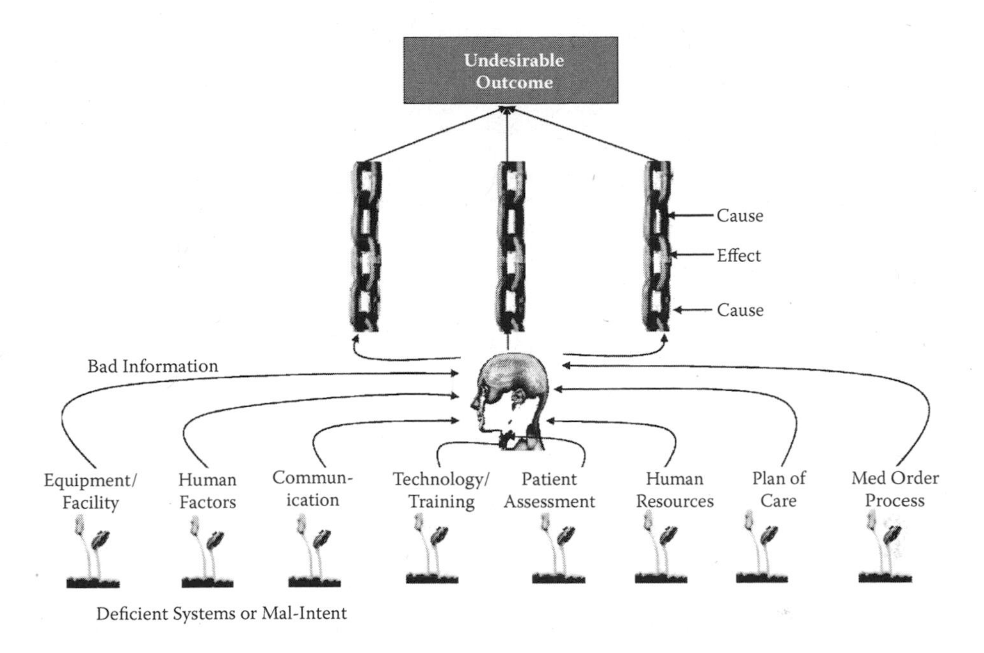

FIGURE 10.3 The germination of a failure.

systems, etc. Organizational systems are put in place to provide users of the systems information to make better decisions. When such systems are obsolete, inadequate, or nonexistent, we increase the opportunity for human error in decision making and therefore undesirable outcomes.

As bad information is fed to an individual, the person must internalize this information along with his or her own training, experience, education, and past successes and determine what actions are appropriate. A poor decision will result in the triggering of a series of observable cause-and-effect relationships that, left unchecked, will eventually lead to an undesirable outcome that will have to be addressed whether we like it or not.

In summary, these organizational systems represent the cause categories we have been describing. Specific causes identified within these categories are referred to as Latent Root Causes. As these deficient systems feed poor information to an individual, the individual is more apt to make a poor decision. We will refer to this decision error as a Human Root Cause. When humans take inappropriate actions, they trigger physical consequences to occur. We will refer to the initial physical consequences as Physical Root Causes. This root system will be described in detail later in this chapter. Understanding the germination of a failure is important before we attempt to try and graphically express it.

CONSTRUCTING A LOGIC TREE

Let's move on and delve into the details of constructing the PROACT RCA tool of choice called a "Logic Tree." This is our means of organizing all the data collected

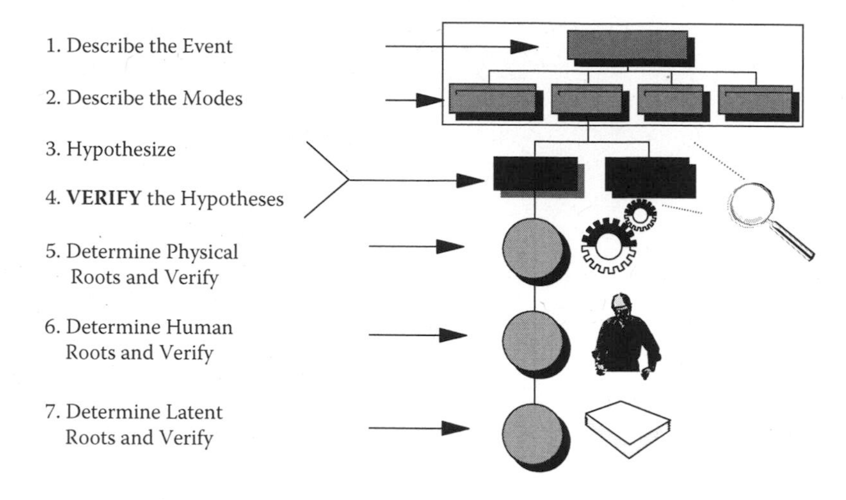

FIGURE 10.4 Logic Tree architecture.

thus far and putting it into an understandable and logical format for comprehension. This is different than the traditional logic diagram and a traditional fault tree. A logic diagram is typically a decision flowchart that will ask a question and, depending on the answer, will direct the user to a predetermined path of logic. Logic diagrams are popular in situations where the logic of a system has been laid out to aid in human decision making. For instance, an operator in a nuclear power generation facility might use such a logic diagram when an abnormal situation arises on the control panel and a quick response is required. A 911 operator might refer to a logic diagram under certain circumstances and ask the caller a series of questions. Based on the caller's answers, the string of questioning would change.

A fault tree is traditionally a totally probabilistic tool that utilizes a graphical tree concept that starts with a hypothetical event. For instance, we may be interested in how an event could occur so we would deduce the possibilities on the next lower level.

A logic tree is a combination of both of the above tools. The answer to certain questions will lead the user to the next lower level. However, the event and its surrounding modes (manifestations) will be factual versus hypothetical. Figure 10.4 shows the basic logic tree architecture. We will begin to dissect this architecture to gain a full understanding of each of its components in order to obtain a full understanding of its power.

The Event

Following is a brief description of the undesirable outcome being analyzed. This is an extremely important block because it sets the stage for the remainder of the analysis. THIS BLOCK MUST BE A FACT. It cannot be an assumption. From an equipment standpoint, the event is typically the loss of function of a piece of equipment and/or process. From a production standpoint, it is the reason that the organization cares about the undesirable outcome. Under certain conditions we will accept such an undesirable outcome, whereas in other conditions we will not.

The *Event* is usually ill defined and there is no standard against which to benchmark, as no common definition exists. Many people believe that they do RCA on incidents. However, if they look back on the ones they have done, they would likely find that they probably were doing RCA because of some type of negative consequence. It is usually negative consequences that trigger an RCA, not necessarily the incident itself. Think about it. I may think I am doing an RCA because a pump failed, but I am really doing it because it stopped production. If the same pump failed and there was not a negative consequence, would I be doing an RCA on the failure?

In a hospital if a patient is given the wrong medication it is called an Adverse Drug Event (ADE). If a patient receives the wrong type, frequency, or dose of a medication but has no adverse side effects, are we going to do a full-blown RCA? If the patient receives the wrong type, frequency, or dose of medication and has an allergic reaction and dies, we will likely conduct a full-blown RCA (or someone will). The point we are trying to make is that the magnitude and severity of the negative consequences will usually dictate whether or not an RCA will be commissioned, and also the depth and breadth of the analysis to be conducted.

When we are in a business environment that is not sold out (meaning we cannot sell all we can make), we are more tolerant of equipment failures that restrict capacity because we do not need the capacity anyway. However, when sales pick up and the additional capacity is needed, we cannot tolerate such stoppages and rate restrictions. In the non-sold-out state, the event may be accepted. In the sold-out state, it is not accepted. This is what we mean by the event being defined as "the reason we care." We only care here because we could have sold the product for a profit.

Please remember our earlier discussion about the Error-Change Phenomenon in the "Preserving Event Data" chapter (Chapter 8). We discussed how "error-change relationships" are synonymous with "cause-effect relationships." The event is essentially the last link in the error chain (see Figure 10.5). It is the last effect and usually how we notice that something is wrong.

The Mode(s)

The modes are a further description of how the Event HAS occurred in the past. REMEMBER, THE EVENT AND MODE LEVELS MUST BE FACTS. This is what separates the logic tree from a fault tree. It is a deduction from the Event block and seeks to break down the bigger picture into smaller, more manageable blocks. Modes are typically easier to delineate when analyzing chronic events. Let's say here that we have a process that continually upsets. We lose production capacity for various reasons (modes). The Top Box in Figure 10.6 would describe this situation (Event plus Mode Level).

In this case the process *has been upset in the past* due to motor failure, pump failure, fan failure, and shaft failure. These modes represent individual occurrences. This does not mean that they do not have common causes, but their occurrences surfaced separately. Essentially the modes are answering the question, "How has the event occurred in the past?"

Event	Mode	Frequency	Impact	Total Lost Units	Total Loss
Pump Failure	Bearing Problems	12	500	6,000	$180,000
Off Spec. Product	Wrong Color	52	400	20,800	$624,000
Conveyor Failures	Roller Failures	500	50	25,000	$750,000

Subsystem	Event	Mode	Frequency	Impact	Total Loss
Subsystem A	Event 1	Mode 11	30	$40,000	$1,200,000
Subsystem A	Event 2	Mode 7	4	$230,000	$920,000
Subsystem B	Event 3	Mode 1	365	$1,350	$492,750
Subsystem A	Event 2	Mode 5	10	$20,000	$200,000
Subsystem A	Event 2	Mode 8	10	$10,000	$100,000
Subsystem B	Event 5	Mode 6	35	$2,500	$87,500
Subsystem B	Event 4	Mode 4	1000	$70	$70,000
Subsystem A	Event 4	Mode 12	8	$8,000	$64,000
Subsystem B	Event 6	Mode 10	6	$8,000	$48,000
Subsystem C	Event 4	Mode 13	4	$7,500	$30,000
Subsystem B	Event 4	Mode 9	10	$2,500	$25,000
Subsystem A	Event 1	Mode 2	12	$2,000	$24,000
Subsystem A	Event 1	Mode 3	9	$2,500	$22,500
Subsystem C	Event 6	Mode 14	6	$3,500	$21,000
Total Loss					$3,304,750
Significant Few Losses (Total Loss × .80)					$2,643,800

Recurring Process Upsets

FIGURE 10.5 Event example.

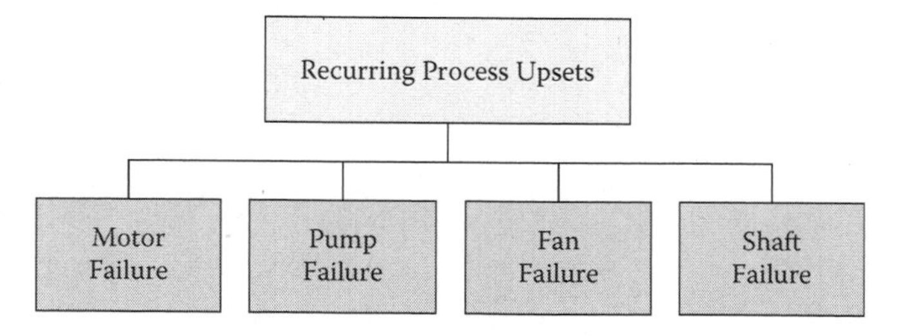

FIGURE 10.6 Top Box example of chronic event.

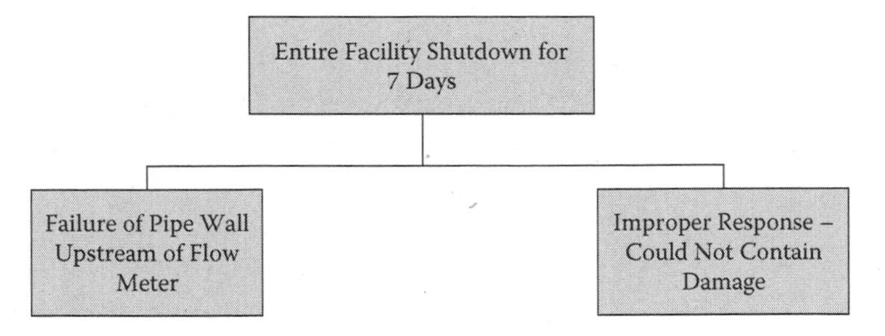

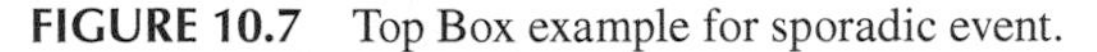

FIGURE 10.7 Top Box example for sporadic event.

When dealing with Sporadic Events (one-time occurrences), we do not have the luxury of repetition so we must rely on the facts at the scene. The modes will represent the manifestation of the failure. The mode will be what triggered the negative consequence to occur. Sporadic failures may have fewer modes than chronic failures. This is because chronic failures represent more than one occurrence. It is not uncommon with a sporadic event to have a single mode (Figure 10.7).

Notice in this case that we included *inadequate response* as a mode. Why? Remember we said earlier that a mode was something that triggered the event to occur. This is also true of when we are looking at what actions after the incident occurred made it worse. Did our response increase the magnitude and severity of the consequences (the Event)? By adding this mode, we will be seeking to identify our response to the incident and uncover when defensive systems were in place at that time and whether or not they were appropriate. If they existed and were appropriate, did we follow them? If we did not follow them, why not? We can see from this line of questioning that we will uncover the system flaws in our response. This way we can implement countermeasures to fill the cracks in our response plans in the future. In our example, the existence of a failed pipe wall along with an improper response caused the initial limited effect of the failure to spread to the entire facility. In this scenario, a 600 lb. steam line ruptured and because the steam lines were not properly contained in a timely manner, all boilers were drained, shutting down the entire facility.

The Top Box

The Top Box is the aggregation of the Event and the Mode levels. As we have emphatically stated, THESE LEVELS MUST BE FACTS! We state this because it has been our experience that the majority of the time we deal with RCA teams, there is a propensity to act on assumptions as if they are facts. This assumption and subsequent action can lead an analysis in a completely wrong direction. The analysis must begin with facts that are verified and conventional wisdom; ignorance and opinion should not be accepted as fact.

To illustrate the dangers of accepting opinion as fact, we will relate a scenario we encountered. We were hired by a natural gas processing firm to determine how to eliminate a phenomenon called "foaming" in an Amine Scrubbing Unit used in their process. In order to get the point across about the Top Box and not have to get into a technical understanding, the illustration in Figure 10.8 is a basic drawing of a scrubbing unit (very similar to a distillation column) with bubble-cap trays and

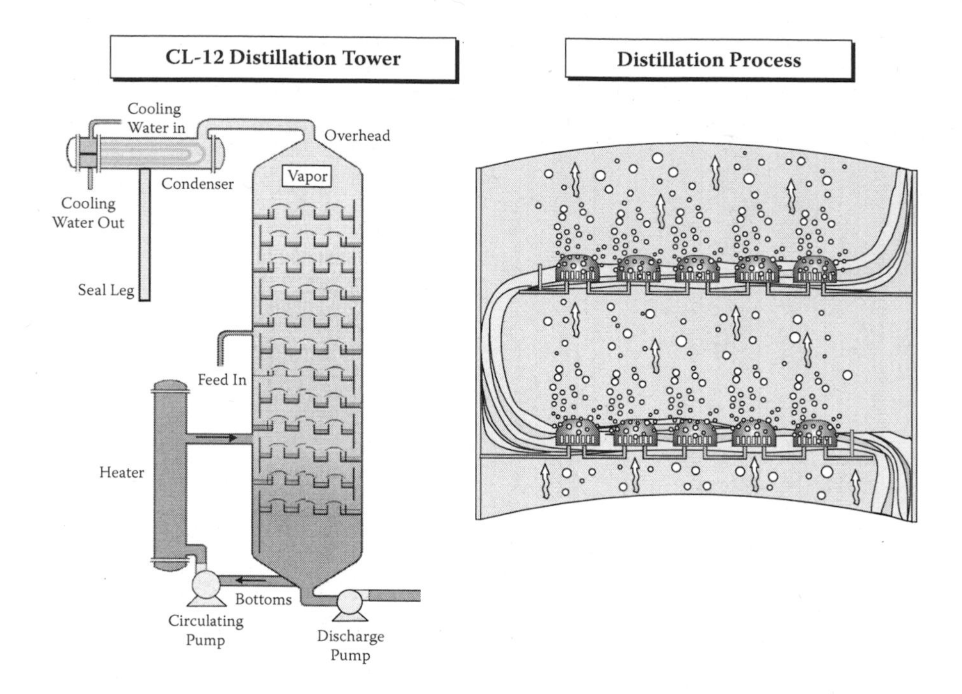

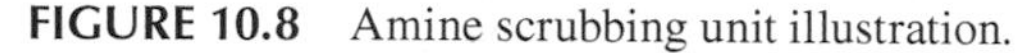
FIGURE 10.8 Amine scrubbing unit illustration.

downcomers. The purposes of this vessel are to clean and sweeten the gas for the gas producers and make it acceptable for the gas producer's customers.

The event described by the company that hired us was vehemently "Foaming." Foaming is a phenomenon by which foam is formed within the Amine Scrubbing Unit, which restricts the flow of gas through the bubble-cap trays. As a result, capacity is restricted and they are unable to meet customer demand due to the unreliability of the process. When asked, "How long has this been occurring?" the reply was 15 years! Why would an organization accept such an event for 15 years? The answer was candid and simple: over the past 15 years there was more capacity than demand. Therefore, rate restrictions were not costing as much money. Now the business environment has changed where demand has outpaced capacity and the facility cannot meet the challenge. Now it is costing them lost profit opportunities.

Given the above scenario, what is the Event and what is the Mode(s)? The natural tendency of the team of experts was to label the event as *foaming*. After all, they had a vested interest in this label as all of their corrective actions to date were geared at eliminating foaming. But what are the facts of the scenario? We were unbiased facilitators of our PROACT RCA methodology; therefore, we could ask any question we wished. What was the real reason they cared about the perceived foaming event? They were only concerned now because they could not meet customer demand. That is the fact! This analysis would not have taken place if the company had not been getting complaints and threats from its customers that they would seek other options.

Given the Event is "Recurring Process Interruptions Prevent Ability to Satisfy Customer Demand," what is the Mode? What is the symptom of why the Event is

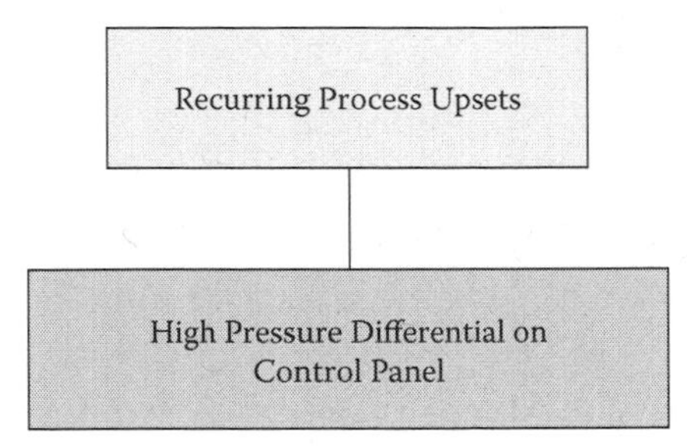

FIGURE 10.9 "Foaming" example of Top Box.

occurring? At this point the natural tendency of the team was to again identify *foaming* as the Mode. Remember, *modes* must be facts. Being unbiased facilitators and not experts in the technical process, we explained that the Amine Scrubbing Unit was a closed vessel. In other words, we could not see inside the vessel to confirm the presence of foam. So we asked, "How do you know you have foam if you cannot see it?" This question seemed to stifle the team for a while as they pondered the answer. Many minutes later (nearly an hour), one of the operators replied that they know they have foaming when they receive a high-pressure differential on the control panel. The instrumentation in the control room was calibrated for accuracy and indicated that the instruments in question were indeed accurate. The FACT in this case was not yet foaming, but a "high pressure differential" on the control panel. This was the indicator that leaped into people's minds, having them believe that foaming did exist. This is a very typical situation where our minds make leaps based on indications. Based on this new information, the Top Box shown in Figure 10.9 was composed.

Coming off of this Mode level, we would begin to hypothesize about how the preceding event could have occurred. Therefore, our question would become, "How could we have a high-pressure differential on the control panel resulting in a restriction of the process?" The answers supplied by the expert team members were Foaming, Fouling, Flooding, and Plugged Coalescing Filters upstream. These were the only causes they could think of for a high-pressure differential on the board. Now came the task of verifying which were true and which were not true.

We simply asked the question, "How can we verify foaming?" Again, a silence overcame the crowd for about 15 seconds until one team member rose and stated that they had taken over 150 samples from the vessel and they could not get any to foam. In essence, they had disproved that foaming existed but would not believe it because it was the only logical explanation at the time. They honestly believed foam was the culprit and acted accordingly.

To make a long story short, the operators received the indicator that a high-pressure differential existed. Per their experience and education they responded to the indicator as a foaming condition. The proper response under the assumed conditions was to shoot into the vessel a liquid called anti-foam, which is designed to break down foam (if it existed). The problem was that no one knew exactly how much anti-foam to shoot in or how much they were shooting in. It turned out the operators were shooting in so much anti-foam that they were flooding the trays. They were treating a condition that did not exist and creating another condition that restricted flow.

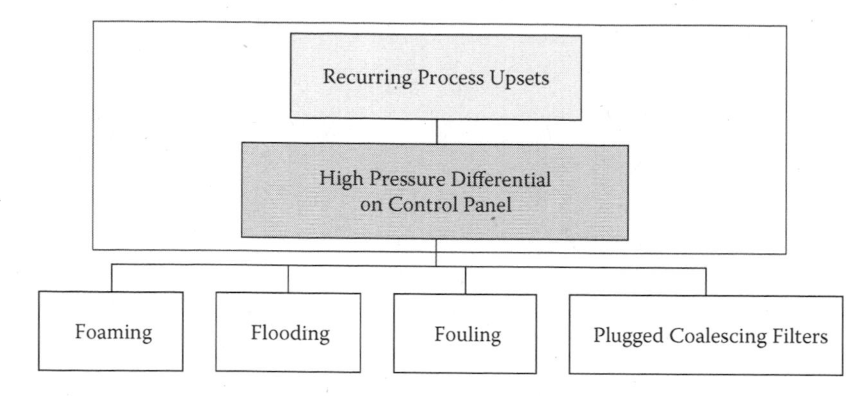

FIGURE 10.10 Foaming Top Box and First Level example.

The original high-pressure differential was being caused by a screen problem in a coalescing filter upstream. But no one ever considered that condition as an option at the time. The point to the whole story is that if we had accepted the team's opinion of *foaming* as fact, we would have pursued a path that was incorrect. This is the reason we are vehement about making sure the Top Box is factual. Figure 10.10 shows how the resulting Top Box and first hypotheses looked.

The Hypotheses

As we learned in school at an early age, a hypothesis is merely an *educated guess*. Without making it any more complex than that, hypotheses are responses to the "How could …" questions described previously. For instance, in the foaming example we concluded that the "high-pressure differential" was the mode. This is the point at which the facts end and we must hypothesize. At this point we do not know why there is a high-pressure differential on the control panel; we just know that it exists. So we simply ask the question, "How could the preceding event have occurred?" The answers we seek should be as broad and all inclusive as possible. As we will show in the remainder of this chapter, this is contrary to normal problem-solving thought processes.

Let's take a few minutes here and discuss the nuances between asking the question "why" as opposed to "how could." Several self-proclaimed RCA techniques involve the use of asking the question *why*. Such tools include the 5 Why's and various types of Why Trees. Rather than get into the pros and cons of the approaches themselves, we will make one key distinction between them and PROACT's Logic Tree tool. When we ask the question *why*, we are connoting two things in our anticipated response: (1) that we seek a singular answer and (2) that we want an individual's opinion. From our standpoint, based on these premises, asking *why* encourages a narrow range of possibilities and allows assumption to potentially serve as fact. If we are seeking someone's opinion without backing it up with evidence, it is an assumption. This allows ignorance to creep into analyses and serve as fact.

On the flip side, what do we seek when asking *how could* something occur? This line of questioning promotes seeking all of the possibilities instead of the most likely.

Keep in mind that the reason chronic events occur is because our conventional thinking has not been able to solve them in the past. Therefore, the true answers lie in something "unlikely" that will be captured by asking *how could* as opposed to *why*. Based on our responses to the *how could* questions, we will tap into our 5P's data that we collected earlier and use it to prove or disprove our hypotheses.

This distinction may seem like semantics, but it is a primary key to the success of any RCA. Only when we explore all the possibilities can we be assured that we have captured all of the culprits. In PROACT RCA what we prove *not* to be true is just as important (if not more important) as what we prove to be true.

Verifications of Hypotheses

As mentioned previously, hypotheses that are accepted without validation are merely assumptions. This approach, though a prevalent problem-solving strategy, is really no more than a *trial-and-error* approach. In other words, it appears to be this case so we will spend money on this fix. When that does not work, we reiterate the process and spend money on the next likely cause. This is an exhaustive and expensive approach to problem solving. Typically, brainstorming techniques such as Fishbone and 5 Why's and troubleshooting approaches do not require validation of hearsay with evidence. This can be dangerous and expensive. Dangerous because all of the causes have not been identified or verified, leaving us open to the risk of recurrence. Expensive because we may keep spending money until something finally works.

In the PROACT RCA methodology all hypotheses must be supported with hard data. The initial data for this purpose was collected in our 5P's effort in the categories of Parts, Position, People, Paper, and Paradigms. The 5P's data will ultimately be used to validate hypotheses on the logic tree. While this is a vigorous approach, the same parallel is used for the police detective preparing for court. The detective seeks a solid case and so do we. A solid case is built on facts, not assumptions. Would we expect a detective to win a murder case based on the sole testimony of a convicted drug dealer. This is a weak case and not likely to be successful.

In the analogy illustrated in Figure 10.11, the Top Box is equated to the crime scene or the facts. When all we have to deal with are the facts, we start to question *how could* these facts exist in this form. The answers to these questions represent our hypotheses. To a criminologist, they represent leads. Leads must be validated with evidence, and all they do is *lead* to asking another question and the process continues. Eventually we will uncover what we call physical causes and what the detectives call forensic evidence. Just like in the television series CSI (Crime Scene Investigation), people who do laboratory forensic work deal with the "hows" or the physical evidence. Their role is not to determine the *whys*.

The *whys* are analogous to motive and opportunity in a criminal investigation. As we will discuss later in this chapter, PROACT associates the terms Human and Latent Root Causes with motive and opportunity. Prosecutors must prove why the defendant chose to take the actions he or she did that triggered the physical evidence to occur and eventually commit a crime.

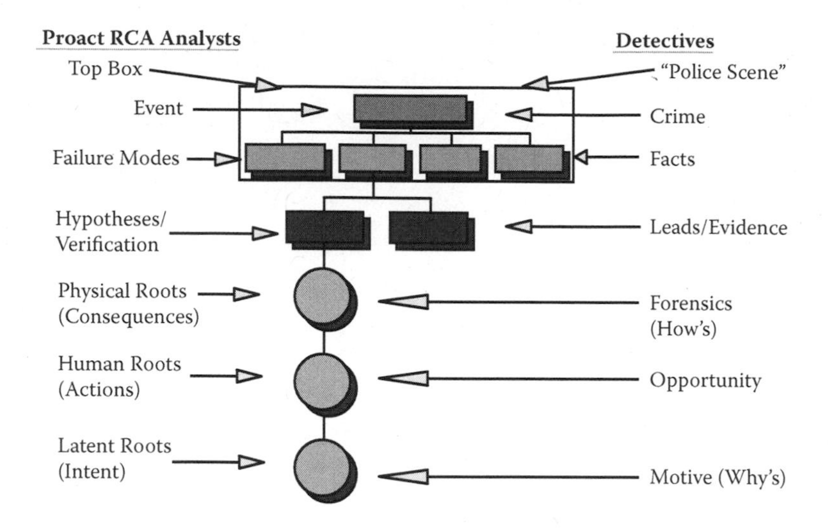

FIGURE 10.11 Logic Tree commonalities between RCA and criminal investigation.

For the purposes of this discussion on verification of hypotheses, we will use the following definition of evidence:

> ***Evidence:*** *Any data used to prove or disprove the validity of a hypothesis in the course of an investigation and/or analysis.*

The literal definition of evidence will mean different things to different people based on their occupations. The meaning of evidence in the eyes of the law may be different than evidence to a root cause analyst. We have defined evidence in the manner above because it is simple, to the point, and represents how we use the term in Root Cause Analysis.

Hard data for validation means eyewitness accounts, statistics, certified tests, inspections, on-line measurement data, and the like. A hypothesis that is proven to be true with hard data becomes a fact.

It is important to note the concept of Cognitive Dissonance related to our interpretation of evidence in an RCA. Cognitive Dissonance is a state of tension that occurs whenever a person holds two cognitions (ideas, beliefs, attitudes, opinions) that are so psychologically inconsistent, such as "smoking is a dumb thing to do because it could kill me" and "I smoke two packs a day."* In layman's terms, we often tend to justify poor decisions despite the evidence presented. Part of our investigation has to be to understand why we overrode the convincing evidence.

Along these same lines, we must also understand the effect of Confirmation Bias. If new information is consistant with our beliefs, we think it is well founded and useful: "Just what I always said." But if the new information is dissonant, then we

* Tavris, Carol and Elliott Aronson. 2007. *Mistakes Were Made (But Not by Me).* Orlando: Harcourt, Inc., p. 12.

Hypothesis	Verification Method	Responsibility	Completion Date	Outcome	Confidence
Foam	Performed Foam Tests on Samples	RHT	09/09/99	FALSE	0
Flooding	Monitor Volume of Anti-Foam in Scrubber	TGJ	09/10/99	TRUE	5
Fouling	Inspect Trays	RPG	09/11/99	FALSE	0
Plugged Coalescing Filters	Inspect Filters	RCA	09/12/99	TRUE	5

FIGURE 10.12 Sample Verification Log.

consider it biased or foolish: "What a dumb argument!"* When we are leading analyses we should be very cognizant of Confirmation Bias and not allow our personal biases to interpret the evidence at hand.

This is especially tempting when dealing with our legal system, and our safety analysis may enter a legal realm. In legal circles, polarization of evidence is more likely than in a safety investigation. In court, we have a plaintiff and a defendant. The plaintiff's charges outline the plaintiff's case and the defense has to counter that position. There are two sides looking at the same set of facts. Each side will interpret the facts to its own benefit (confirmation bias). Evidence that is entered that agrees with our case, we readily accept. Evidence that contradicts our case, we seek to discredit.

In keeping with our "solid case" analogy, we must keep in mind that organization is a key to preparing our case. To that end, we should maintain a Verification Log on a continual basis to document our supporting data. Figure 10.12 provides a sample of a Verification Log used in the PROACT RCA methodology. This document supports the logic tree and allows it to stand up (especially in court).

The Fact Line

The fact line starts below the *mode* level because above it are facts and below it are hypotheses. As hypotheses are proven to be true with hard data and become facts, the fact line moves down the length of the tree. For instance, Figure 10.13 illustrates this for the case of the foaming example mentioned earlier.

Physical Root Causes

The first root level causes that are encountered through the reiterative process will be the Physical Roots. Physical Roots are the tangible roots or component-level

* Tavris, Carol and Elliott Aronson. 2007. *Mistakes Were Made (But Not by Me)*. Orlando: Harcourt, Inc., p. 18.

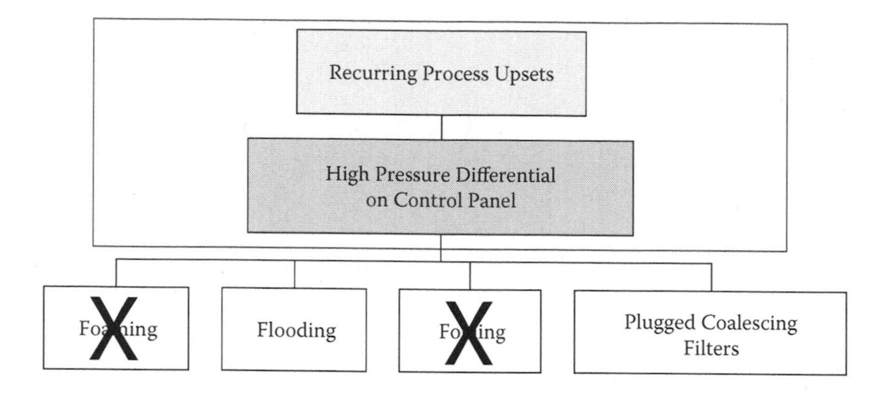

FIGURE 10.13 The fact line positioning.

roots. Physical Roots are observable. In many cases, when undisciplined problem-solving methods are used, people will have a tendency to stop at this level and call them "Root Causes." We do not subscribe to this type of thinking. In any event, all Physical Root causes must also endure validation to prove them as facts. Physical Roots are generally identifiable on the logic tree by the fact that they are usually the first perceptible consequences after a human decision error has been made. In terms of logic tree orientation, Physical Roots generally are located shortly after the human roots (decision error roots) have been identified.

Human Root Causes

Human Root causes will almost always trigger a Physical Root cause to occur. Human Root causes are decision errors. These are either errors of omission or commission. This means that either we decided not to do something we should have done, or we did something we were not supposed to do. Examples of errors of omission might be that we were so inundated with reactive work, we purposely put needed inspection work on the back burner to handle the failures of the day. An error of commission might be that we aligned a piece of equipment improperly because we did not know how to do it correctly.

Human Root causes are not intended to represent the vaguely used term of Human Error. We use the Human Root only to represent a human decision that triggered a series of physical consequences to occur. In the end, this series of physical consequences ultimately resulted in an undesirable outcome. Ending an analysis with a conclusion of "Human Error" is a cop out. It is vague and usually indicates that it is not knows why the incident occurred. Oftentimes, we as the public are told that airplane accident investigations result in pilot error. This should be offensive to the general public because that pilot's life depended on the decision that he made and he knew that. Therefore, we can reasonably conclude that he was making the best decisions he could at the time. What does that tell us? It can tell us many things, some of which include (1) the pilot was not trained properly for the situation he or she encountered, (2) the procedure the pilot did follow was inadequate for some reason, (3) the pilot did not follow the appropriate procedures (in which we would have to ask *why*), or the pilot was provided poor information from either the instrumentation or air traffic control. To simply say Human Error does not describe what actually happened.

PROACT seeks to uncover the reasons people thought they were making the right decision at the time they made the decision. We refer to this basis of the decision as the Latent Roots. These are the traps that result in poor decisions being made.

The frame of reference for understanding people's behavior and judging whether it made sense is in their own normal work context, the context in which they were embedded. This is the point of view from where decisions assessments are sensible, normal, daily, unremarkable, and expected. The challenge, if we really want to know whether people anticipated risks correctly, is to see the world through their eyes, without knowledge of outcome, without knowing exactly which piece of data will turn out to be critical afterward.*

As we discussed in detail in Chapter 7 ("The Role of Human Error in Root Cause Analysis"), this is the point in the logic tree where the top 10 error contributors are explored in depth:

1. Ineffective Supervision
2. Lacking Accountability System
3. Distractive Environment
   a. Low Alertness
   b. Complacency
4. Work Stress/Time Pressure
5. Overconfidence
6. First-Time Task Management
7. Imprecise Communications
8. Vague or Incorrect Guidance
9. Training Deficiencies
10. New Technology

Each of these error traps could be a text by itself. The point we wanted to make here is that by understanding the conditions that increase the risk of human error in decision making, we can implement proactive changes to reduce the risk. We are not perfect beings, so we will never eliminate human error in decision making. The misattribution of errors is one reason we fail to learn from our mistakes: we haven't understood their root causes.† But that does not mean we cannot strive for such perfection, as success will be achieved during the journey.

While the questioning process thus far has been consistent with asking *how could*, at the Human Root level (decision error) we want to switch the questioning to "why?" When dealing in the physical and process areas, we cannot ask equipment *why* it failed. Only at the Human Root level do we encounter a person's involvement. When we get to this level, we are not interested in *whodunit*, but rather why they made

* Dekker, Sidney. 2007. *Just Culture: Balancing Safety and Accountability*. Hampshire, England: Ashgate Publishing, p. 72.

† Hallinan, Joseph T. 2009. *Why We Make Mistakes*. New York: The Doubleday Publishing Group, p. 189.

the decision that they did at the time they did. Understanding the rationale behind decisions that result in error is the key to conducting true RCA. Anyone who stops an RCA at the Human Level and disciplines an identified person or group is participating in a witch-hunt. Witch-hunts were discussed in the Preserving Failure Data section and proved to be non-value added, as the true roots cannot be attained in this manner. This is because if we search for a scapegoat, no one else will participate in the analysis for fear of repercussions. When we cannot find out why people make the decisions they do, we cannot permanently solve the issue at hand. Therefore, we cannot eliminate its risk of recurrence.

Latent Root Causes

Latent Root causes are the organizational systems that people utilize to make decisions. When such systems are flawed, they result in decision errors. The term "Latent"* is defined as

> ***Latent:*** *Those adverse consequences that may lie dormant within the system for a long time, only becoming evident when they combine with other factors to breach the system's defenses.*

When we use the term organizational or management systems, we are referring to the *rules and laws* that govern a facility. Examples of organizational systems might include policies, operating procedures, maintenance procedures, purchasing practices, stores and inventory practices, training systems, quality control mechanisms, etc. These systems are all put in place to help people make better decisions. When a system is inadequate or obsolete, people end up making decision errors based on flawed information. These are the true root causes of undesirable events. We have now defined the most relevant terms associated with the construction of a logic tree. Now let's explore the physical building of the tree and the thought processes that go on in the human mind.

Experts who participate on such teams are generally well-educated individuals, well respected within the organization as problem solvers, and people who pay meticulous attention to detail. With all this said and done, using the logic tree format, an expert's thought process may look like the Figure 10.14.

This poses a potential hurdle to a team's success, because for the most part, the analysis portion is bypassed and we go straight from problem definition to cause. It is the Principal Analyst's responsibility to funnel the expertise of the team in a constructive manner without alienating the team members. Such an RCA team will have a tendency to go to the *micro* view and not the *macro* view. However, in order to understand exactly what is happening, we must step back and look at the big picture. In order to do this, we must derive exactly where our thought process originated from and search for assumptions in the logic.

A logic tree is merely a graphical expression of what a thought would look like if it were on paper. It is actually looking at how we think. Let's take a simple example

* Reason, James. 1990–1992. *Human Error.* Victoria: Cambridge University Press, p. 173.

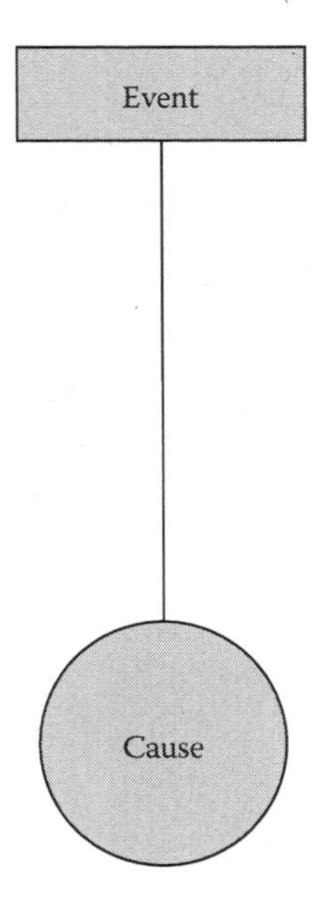

FIGURE 10.14 The expert's Logic Tree.

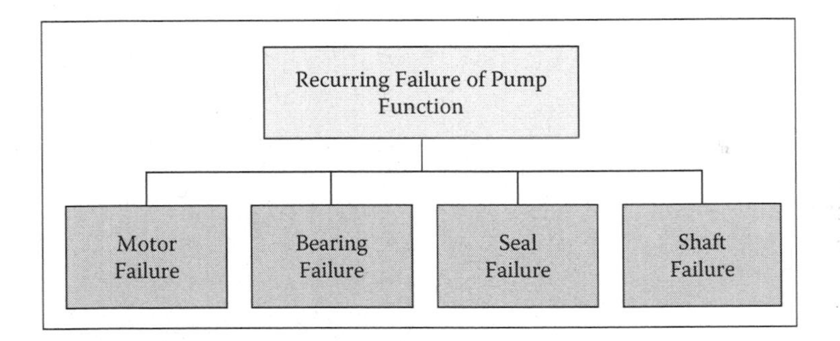

FIGURE 10.15 Recurring pump failure example.

of a pump of some type that is failing. We find that 80% of the time this pump is failing due to a bearing failure. This shall serve as the *mode* that we pursue first for demonstration purposes (Figure 10.15).

BREADTH AND ALL-INCLUSIVENESS

If we have a team of operations, maintenance, and technical members and ask them the question, "How could a bearing fail?" their answers would likely get into the nuts and bolts of such details as improper installation, design error, defective materials, too much or too little lubricant, misalignment, and the like. While these are all very valid, they jump into too much detail too fast. We want to use deductive logic in short leaps.

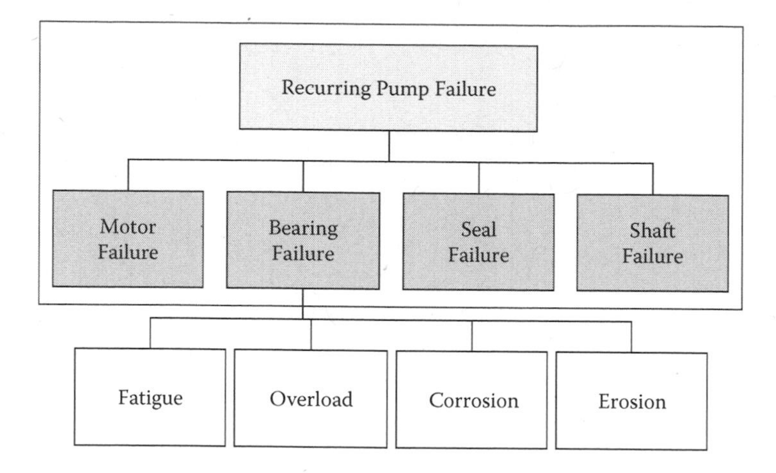

FIGURE 10.16 Broad and all-inclusive thinking.

In order to be *broad and all-inclusive* at each level, we want to identify all the possible hypotheses in the fewest blocks (Figure 10.16). To do this, we must imagine we are the part being analyzed. For instance, in the above example with the bearing, if we thought of ourselves as being the bearing, we would think, "How exactly did we fail?" From a physical failure standpoint, the bearing would have to erode, corrode, overload, or fatigue. These are the only ways the bearing can fail. All of the hypotheses developed earlier by the experts (the micro answers) would cause one or more of these failure states to occur.

From this point we would have a metallurgical review of the bearing conducted. If the results were to come back and state that the bearing failed due to fatigue, then there are only certain conditions that can cause a fatigue failure to occur. The data or evidence leads us in the correct direction, not the team leader. This process is entirely data driven.

If we are broad and all-inclusive at each level of the logic tree and we verify each hypothesis with hard data, then the fact line drops until we have uncovered all the root causes. This is very similar in concept to many Quality initiatives. The more popular Quality initiatives focused on Quality of the entire manufacturing process instead of just checking Quality of the finished product (when it was too late).

THE ERROR-CHANGE PHENOMENON APPLIED TO THE LOGIC TREE

Now let's explore how the error-change concept (cause-effect relationship) parallels the logic tree. As we explore the path of the logic tree, there are three key signs of hope in favor of our finding the true root causes. These keys are as follows (Figure 10.17):

1. Order
2. Determinism
3. Discoverability

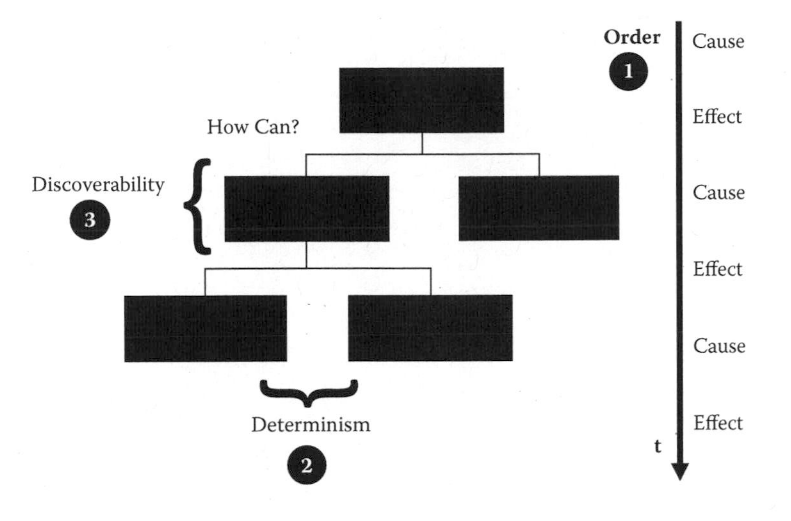

FIGURE 10.17 The three keys.

Order

If we truly believe that the error-change phenomenon exists, then we have the hope that following cause-and-effect relationships backward will lead us to the culprits; the root causes. We often ask our classes if they believe there is *order* in everything, including nature. There is generally a silent pause until they think about it and they cite facts such as tides coming in and going out at predetermined times, the sun rising and setting at predetermined times, and the seasons that various geographic regions experience on a cyclical basis. These are all indications that such order, or patterns, exists.

Determinism

This means everything is determinable or predictable within a range. If we know a bearing has failed, the reasons (hypotheses) of how the bearing can fail are determinable. We discussed this earlier with the options being corrosion, erosion, fatigue, and overload. This is determinable within a range of possibilities.

People are the same way, to a degree. People's behavior is determinable within a broader range than equipment because of the variability of the human race. If we subject humans to specific stimuli, they will react within a certain range of behaviors. If we alienate employees publicly, chances are they will withdraw their ability to add value to their work. They in essence become human robots because we treated them that way.

Determinism is important because when constructing the logic tree it becomes essential, from level to level, to develop hypotheses based on determinism.

Discoverability

This is the simple concept that when you answer a question it merely begets another question. We like to use the analogy of children in the age ranges of 3 to 5 years old. They make beautiful Principal Analysts because of their inquisitiveness and openness to new information. We have all experienced our children at this age when they say, "Daddy, why does this happen?" We can generally answer the series of *why* questions about five times before we do not know the actual answer. This is discoverability: questions only lead to more questions. On the logic tree, discoverability is expressed from level to level when we ask, "How could something occur?" and the answer only leads to another *how could* question.

FINDING PATTERN IN THE CHAOS

All of these keys provide the analyst the hope that there is a light at the end of the tunnel and it is not a train. We are basically searching for pattern in a sea of chaos and these keys help us find pattern in the chaos.

Imagine if we were the investigators at the first bombing site of the Twin Towers buildings in New York back in 1993. Could we even visualize finding the answer from looking at the rubble generated from the blast, the chaos? Yet, the investigators knew that there was a pattern in the chaos somewhere and they were going to find it. Apparently, within 2 weeks of that blast, the investigators knew the type of vehicle, the rental truck agency, and the make-up of the bomb. This is true faith in finding pattern in chaos. These people believe in the logic of failure.

AN ACADEMIC EXAMPLE

Let's take all of the described pieces of the logic tree architecture and put them into perspective in an academic example to which we can all relate.

We have all experienced problems, at some time or another, with our Local Area Networks (LANs) at our offices. This is a very universal issue we see in our travels. If we were to look at this issue from an RCA perspective, what is the Event in this case? The end of the error chain is that the LAN is not functioning as it was designed and in a manner to which we are accustomed. Therefore, we may want to paraphrase and say, "Recurring LAN Failures" is our Event because it is the reason we care—the last effect of the cause and effect chain (Figure 10.18).

Subsystem	Event	Mode	Frequency	Impact
Recovery	Recirculation Pump Fails	Bearing Locks Up	12	12 hours
Recovery	Recirculation Pump Fails	Oil Contamination	6	1 day
Recovery	Recirculation Pump Fails	Bearing Fails	12	12 hours
Recovery	Recirculation Pump Fails	Shaft Fracture	1	5 days
Subsystem	**Event**	**Mode**	**Frequency**	**Impact**
Recovery	Recirculation Pump Fails	Bearing Problems	12	12 hours
Recovery	Recirculation Pump Fails	Shaft Fracture	1	5 days

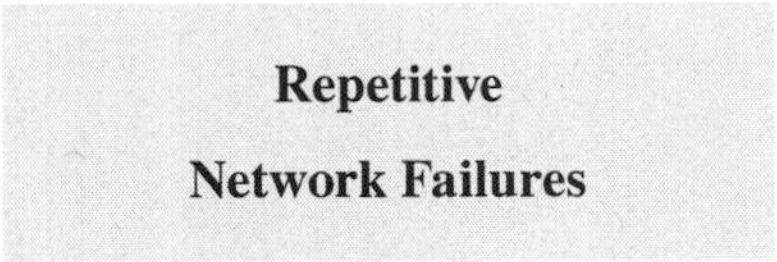

Repetitive

Network Failures

FIGURE 10.18 The LAN Event Block.

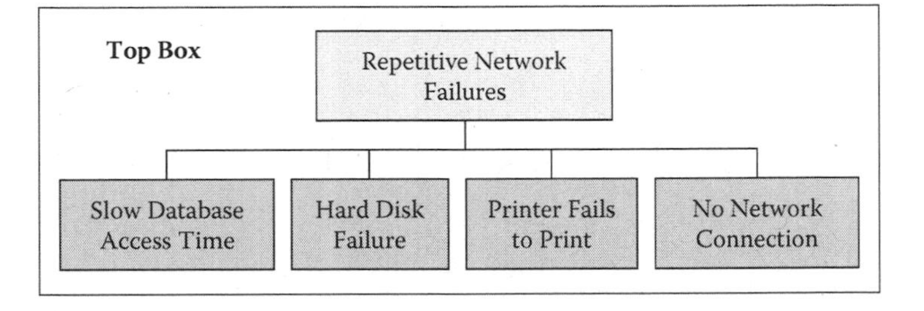

FIGURE 10.19 LAN example Top Box.

Now let's move to the second level and describe the Modes that are the indicators of how we know that the LAN has not been performing as designed. This information, in this situation, may come from users at their workstations in the form of complaints to the Information Systems (IS) department. Some examples of Modes at this level might be Slow Database Access Time, Hard Disk Failure, Printer Fails to Print, and No Network Connection. These are all facts, being that the users have observed them in the past.

Which Mode would we want to approach first? Had we performed an Opportunity Analysis (OA) type of assessment we would already have the Event and the Mode that has been the most costly to the organization. In this example we are going to pursue the Mode with the greatest impact as reported by the users and that would be that the "Printer Fails to Print." In this particular office, the majority of the complaints have been that the printer does not print when they send a job to it. These complaints absorb about 80% of the IT technician's time. For this reason, we will pursue this leg. Figure 10.19 shows what the Top Box might look like.

At this point we begin our hypothetical questioning into, "How could the printer fail to print?" The natural thought process would be to respond with such answers as no toner, no power, wrong configuration, operator error, no paper, etc. (the micro answers). All of these hypotheses are valid, but do they meet the criteria of *broad and all-inclusive*? This is the most difficult portion of constructing a Logic Tree, thinking broadly! Do all of the possibilities exist somewhere in the printer, the computer, the cable, or the operator? The next level of this logic tree would look something like Figure 10.20.

Now comes the task of proving which hypotheses are true and which are not. It is at this point that the Verification Log begins to be developed and we utilize information collected in our 5P's as validation data. Let's take the first hypothesis of the printer and determine a test that can prove or disprove it. We can take our laptop

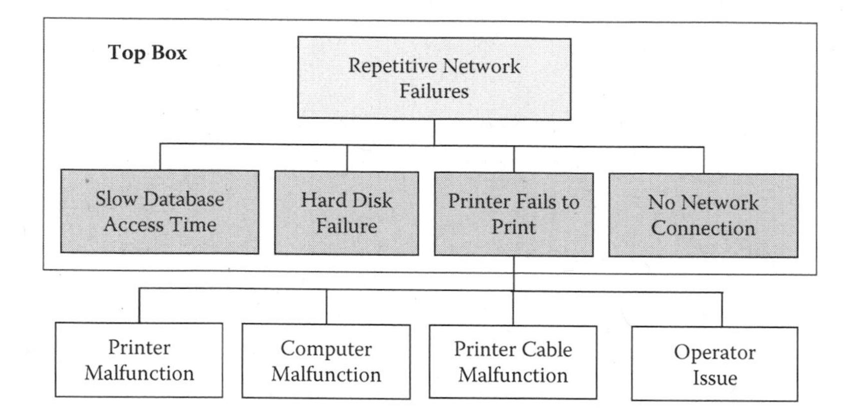

FIGURE 10.20 The first hypothetical leg.

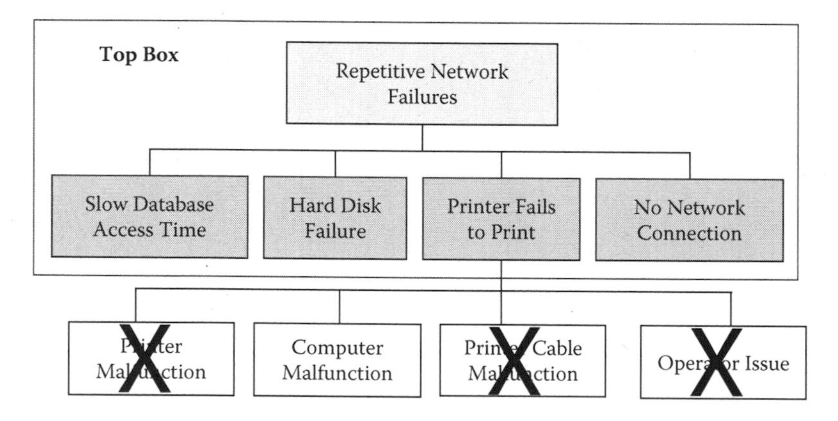

FIGURE 10.21 Updated Logic Tree.

computer and connect it to this printer with the same cable and the same operator to test its functionality. In this case, the printer functions as designed. Based on this test, we can cross out the hypotheses of the printer, the cable, and the operator. However, we cannot select the computer by process of elimination. The computer must also have a test to validate it. In this case we can connect another known working printer to the same computer to test its functionality with the same operator. We conclude from this test that the new printer also does not perform with the same computer. Based on these tests the Logic Tree would look like Figure 10.21.

At this point our *fact line* has moved down from the Mode level to the first hypothesis level. Because the hypothesis of the computer has been verified as *true*, it is now a fact and the fact line drops.

We continue our questioning by pursuing the next level and asking, "How could a computer malfunction causing the printer not to print?" This is the discovery portion of the logic tree where one question only begets another question. Again, a hundred reasons could be thought of as to how a computer could malfunction, but we need to think broadly. The broadest blocks we can think of are "Hardware Malfunction"

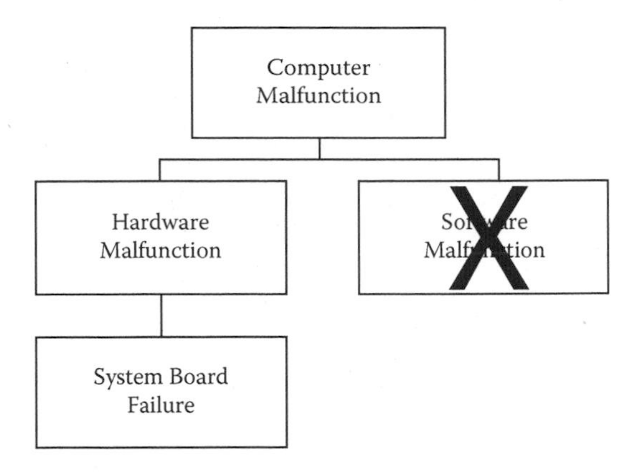

FIGURE 10.22 Hypothesis validation.

and/or "Software Malfunction." Tests must now be developed to prove or disprove these hypotheses. Running diagnostic software determines that the system is reporting an anomaly with the system board, particularly the component that sends instructions to the printer. Other than the identified hardware malfunction, there are no indications of any software malfunctions. This allows us to cross out software malfunction and continue to pursue hardware malfunction (Figure 10.22).

The reiterative questioning continues with, "How could we have a hardware malfunction that would create a computer malfunction that would not allow the printer to print?" We notice in this questioning that we are always reading the logic path back a few levels to maintain the "story" or error path and put the string into the proper context. This helps the team follow the logic tree and put the question into proper perspective. Our broad and all-inclusive answer here, given today's advanced technologies, is a System Board Failure. The previous test of running the diagnostic software confirms an issue with the component in the system board that issues instructions for the printer to print.

At this point in the analysis we removed the system board, cleaned the contact areas, and reseated it properly, making sure the contact is being made and that improper installation concerns are not an issue. The printer still fails to print even after the system board has been installed correctly. Next we replace the system board with one known to be working and properly install it into the computer. This time the printer works as desired. Many feel the analysis is complete at this level because the event will not recur immediately. However, this is the point when we consider the cause to have a *physical root* when the event temporarily goes away. For this reason, we would circle the hypothesis block identified as a *system board failure* indicating it as a physical root cause (Figure 10.23).

Having identified the Physical Root in this case means that we have more work to do in order to uncover the Latent Roots. Our questioning continues with, "How could we have a system board failure that is causing a hardware malfunction that is causing a computer malfunction that is causing the printer not to print?" Either we "Installed It Improperly" and/or we "Purchased It in a Failed State" and/or "It Failed While in Our Use."

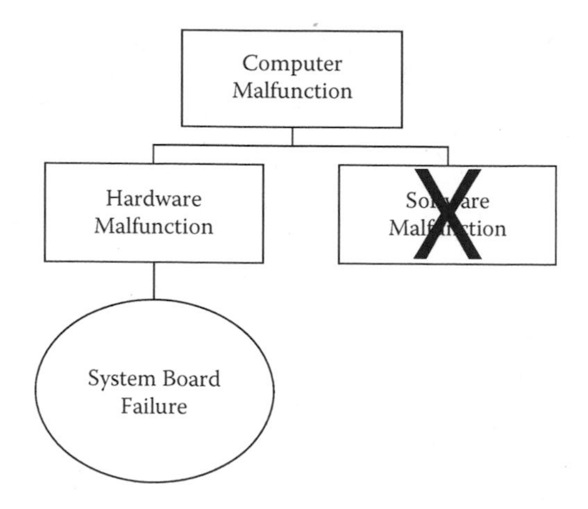

FIGURE 10.23 Identification of Physical Root Cause.

We have already determined that the installation was not an issue. We can eliminate the possibility that the board failed in our possession because interviews reveal that this was a new printer being added to the network and it never worked from the beginning. People on the network therefore chose to divert to another network printer. So this was not a case where the printer worked at one time and then did not work. This serves as proof that the system board did not fail after it ran properly for a period of time, but rather we received it that way from the manufacturer/vendor.

Now we must review our purchasing practices and determine if we have purchasing procedural flaws that allow defective parts to enter the organization. From our 5P's information we determine that there is "No List of Qualified Vendors" and we have "Inadequate Component Specifications." We find that the primary concern for purchasing is to buy based on low cost because that is where the incentives are placed for the purchasing agents in this firm. We have now confirmed that we "Purchased a Poor Quality Card" and because this task involves a conscious human decision that results in an action to be taken, this is deemed our Human Root cause. We circle this block now designating it as such. Remember that at this Human Root level, our questioning switches to *why* because we have reached a human being who can respond to the question. We are now interested in understanding why this individual thought it was the right decision to purchase in this manner at that time. Our question at this point would read, "Why did we choose to purchase this particular system board from another vendor?" Our answers are "No List of Qualified Vendors" and "No Component Specifications" and "Misplaced Incentives." These are the reasons behind the decision and therefore are the Latent Roots (Figure 10.24).

Looking at this example, although academic, could it relate to situations in our own environment where disruptions are caused in our processes due to the infiltration of defective parts into the organization? Without a structured RCA approach, we would utilize trial and error approaches until something worked. This can be very expensive.

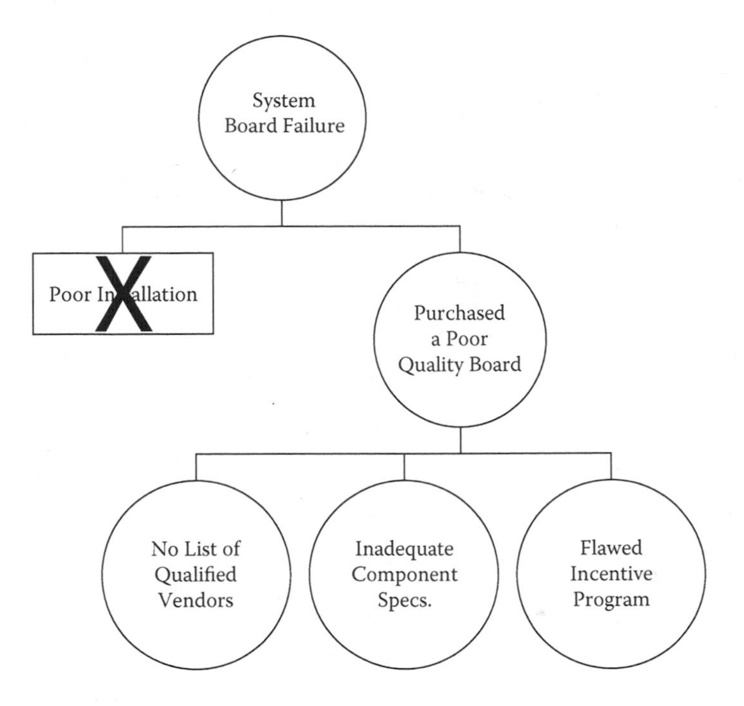

FIGURE 10.24 Identification of Human and Latent Roots.

What if we stopped at the Physical Root of "System Board Failure" and just replaced the board? Would the event likely recur? Sure it would, if the same purchasing habits continue. It may not happen in the same location because not all boards would be received in a failed state, but it would likely happen somewhere else in the organization, forcing another need to analyze.

What if we stopped at the Human Root of "Purchased a Poor Quality Board" and disciplined the purchasing agent who made the decision? Would that prevent recurrence? Not likely, because the decision-making system the agent used is likely being used by other purchasing agents in the organization. It might prevent that agent from making such a decision in the future, but it would not stop other such decisions from being made in the future by other agents.

The only way to prevent recurrence of this event in the entire organization is to correct the decision-making systems we refer to as the Latent Roots or Organizational System Deficiencies. When such deficiencies are uncovered we are truly performing *Root* Cause Analysis.

The completed Verification Log for the above example might look like Table 10.1.

VERIFICATION TECHNIQUES

While we used simple verification techniques in the previous example, there are thousands of ways in which to validate hypotheses. They are all, obviously, dependent

TABLE 10.1
Completed Sample of Verification Log

Hypothesis	Verification Method	Responsibility	Completion Date	Outcome	Confidence
Printer Malfunction	We utilized a stand alone laptop to test the printer.	RMB	09/01/05	Printer worked fine on alternate computer.	0
Computer Malfunction	Utilize a known working printer and test in on the computer.	TDF	09/02/05	Still could not print after the test.	5
Printer Cable Malfunction	We utilized a stand alone laptop to test the cable.	RMB	09/03/05	Cable worked fine on alternate computer.	0
Operator Issue	Have same operator perform same functions in test with new computer and new printer.	RMB	09/04/05	Operator performance not a contributing factor to printer not printing.	0
Hardware Malfunction	Run diagnostic software to check hardware.	TDF	09/05/05	Determined a possible problem with parallel port card.	5
Software Malfunction	Check drivers and configuration.	TDF	09/06/05	Configuration and drivers were correct.	0
Systemboard Failure	Call in a technician to test the systemboard for faults.	TDF	09/07/05	Not indication of systemboard failure.	0
Parallel Port Card Failure	Replace the card with a known working card.	RMB	09/08/05	The document printed fine using the alternate card.	5
Poor Installation	Check installation notes as well as talking with technician who installed the card.	RMB	09/09/05	Installation looked adequate.	0
Purchased a Poor Quality Card	Talk with the purchasing department and storeroom personnel.	RMB	09/10/05	Determined that this was a new installation and discovered the card never worked properly.	5
No List of Qualified Vendors	Determine current vendor requirements.	JCF	09/11/05	Records determined that we have no list of qualified vendors.	5
Flawed Incentive System	Look for a history of low bidder mentality.	FRD	09/12/05	This has been the prevalent purchasing practice in the purchasing department.	5
Inadequate Component Specifications	Check the component specifications for this and other related items.	FGH	09/04/99	We did not have solid component specifications.	5

on the nature of the hypothesis. The following is a list of some common verification techniques used in industrial settings:

- Human Observation
- Fractology
- High-Speed Photography
- Video Cameras
- Laser Alignment
- Vibration Monitoring and Analysis
- Ultrasonics
- Eddy Current Testing
- Infrared Thermography
- Ferrography
- Scanning Electron Microscopy
- Metallurgical Analysis
- Chemical Analysis
- Statistical Analysis (correlation, regression, Weibull Analysis, etc.)
- Operating Deflection (OD) Shapes
- Finite Element Analysis (FEA) Modeling
- Motor Circuit Analysis
- Modal Analysis
- Experimental Stress Analysis
- Rotor Dynamics Analysis
- Capacity and Availability Assessment Program (CAAP®)*
- Work Sampling
- Task Analysis

These are just a few techniques to give you a feel for the breadth of verification techniques that are available. There are literally thousands more. Each of these topics could be a text in itself as well. Many texts are currently available to provide more in-depth knowledge on each of these techniques. However, the focus of this text is on the PROACT RCA methodology. A good Principal Analyst does not necessarily have to be an expert in any or all of these techniques; rather, he or she should be resourceful enough to know when to use which technique and how to obtain the resources to complete the test. Principal Analysts should have a repository of resources they can tap into when the situation permits.

CONFIDENCE FACTORS

It has been our experience that the timelier and more pertinent the data that is collected with regard to a specific event, the quicker the analysis is completed and the more accurate the results are. Conversely, the less data we have initially, the longer the analysis takes and the greater the risk of the wrong cause(s) being identified.

---

* CAAP is a registered trademark of Applied Reliability Incorporated (ARI), Baton Rouge, LA, 1998.

We utilize a *confidence factor* rating for each hypothesis to evaluate how confident we are with the validity of the test and the accuracy of the conclusion. The scale is basic and runs from 0 to 5. A "0" means that without a doubt, with 100% certainty, based on the data collected the hypothesis is *NOT* true. On the flip side, a "5" means that based on the data collected and the tests performed, there is 100% certainty that the hypothesis *IS* true. Between the "0" and the "5' are the shades of gray where the data used was not absolutely conclusive. This is not uncommon in situations where an RCA is commissioned weeks after the event occurred and little or no data from the scene was collected. Also, in catastrophic explosions, we have seen that uncertainty resides in the physical environment prior to the explosion. What formed the combustible environment? These are just a few circumstances in which absolute certainty cannot be attained. The confidence factor rating communicates this level of certainty and can guide corrective action decisions.

We use the rule of thumb that a confidence factor rating of "3" or higher is treated as if it did happen and pursue the logic leg. Any confidence factor rating of less than "3" we treat as a low probability of occurrence and feel it should not be pursued at this time. However, the only hypotheses that are crossed out on the logic tree are the ones that have a confidence factor rating of "0". A "1" cannot be crossed out because it still had a probability of occurring even if the probability was low.

THE TROUBLESHOOTING FLOW DIAGRAM

Once the Logic Tree is completed it should serve as a troubleshooting flow diagram for the organization. Chances are the root causes identified in this RCA will affect the rest of the organization. Therefore, some recommendations will be implemented site-wide or corporation-wide. To optimize the use of a world-class RCA effort, the goal should be the development of a dynamic troubleshooting flow diagram repository. This will end up containing logic diagrams or knowledge management templates that capture the expertise on the organization's best problem solvers on paper. In the introductory chapter we referred to this as *corporate memory*.

Such logic diagrams can be stored on the company's intranet and be made available to all facilities that have similar operations and can learn from the work done at one site. These logic diagrams are complete with test procedures for each hypothesis. They are dynamic because where one RCA team may not have followed one particular hypothesis (because it was not true in their case), the hypothesis may be true in another case and the new RCA team can pick up from that point and explore the new logic path.

The goal of the organization should be to capture the intellectual capital of the workforce and make it available for all from which to learn. This optimizes the intellectual capital of the organization through RCA.

11 Communicating Findings and Recommendations

THE RECOMMENDATION ACCEPTANCE CRITERIA

Let's assume at this point that the complete RCA process has been followed to the letter. We have conducted our modified FMEA and determined our "Significant Few." We have chosen a specific significant event and proceeded through the PROACT process. An identified RCA team has undertaken an organized data collection effort. The team's charter and Critical Success Factors (CSFs) have been determined and a Principal Analyst (PA) has been named. A logic tree has been developed where all hypotheses have been either proven or disproved with hard data. Physical, Human, and Latent roots have been identified. Are we done?

Not quite! Success can be defined in many ways, but an RCA should not be deemed successful unless something has improved as a result of implementing recommendations from the RCA. Merely conducting an excellent RCA does not produce results. As many of you can attest, getting something done about your findings can be the most difficult part of the analysis. Oftentimes recommendations will fall on deaf ears and the entire effort was a waste of your time and the company's money.

If we know that such hurdles will be evident, then we can also proactively plan for their occurrence. To that end, we suggest the development of "Recommendation Acceptance Criteria." We have all faced situations where we spend hours and sometimes weeks and months developing recommendations as a result of various projects only to have the recommendations turned down flat. Sometimes explanations are given and sometimes they are not. Regardless, it is a frustrating experience and it does not encourage creativity in making recommendations. We usually will become more conservative in our recommendations to merely get by.

A Recommendation Acceptance Criteria is what we call the "rules of the game." Managers and executives handle the company's money and, in doing so, make economic decisions as to how the money is spent. In other words, they are the decision makers. Whether these rules are written or unwritten, they define whether or not our recommendations will fly with management. We suggest asking the approving decision makers before we even begin to write our recommendations for "the rules of the game." This is a reasonable request that seeks only to not waste company time and money on non-value-added work.

A sample listing of Recommendation Acceptance Criteria might look like this—the Recommendation must

1. *Eliminate or reduce the impact of the cause*—The goal of an RCA may not always be to eliminate a cause. For instance, if we find our scheduled shutdowns to be excessive, it would not be feasible to expect that they can be eliminated. Our goal may be to reduce the shutdown lengths (Mean-Time-to-Restore [MTTR]) and increase the time between their occurrences (Mean-Time-Between-Failure [MTBF]).
2. *Provide a ___% Return on Investment (ROI)*—Most every company we have ever dealt with has a predetermined ROI. Ten years ago such ROIs were frequently around 15% to 20%. Recently, these expectations have increased dramatically. It is not uncommon to see these numbers in the range of 50% to 100%. This indicates a risk-aversive culture where we only deal with certainty.
3. *Not conflict with capital projects already scheduled*—Sometimes we develop lengthy recommendations only to find that some plans are on the books, unbeknownst to us, that call for the mothballing of a unit, area, or activity. If we are informed of such "secret" plans, then we will not spin our wheels developing recommendations that do not have a chance.
4. *List all the resources and cost justifications*—Decision makers generally like to know that we have thought a great deal about how to execute the recommendations. Therefore, cost/benefit analyses, manpower resources required, materials necessary, safety and quality considerations, etc., should all be laid out.
5. *Have a synergistic effect on the entire system/process*—Sometimes in our working environments we have "kingdoms" that develop internally and we end up in a situation where we stifle communication and compete against each other. This scenario is common and counterproductive. Decision makers should expect recommendations that are synergistic for the entire organization. Recommendations should not be accepted if they make one area look good at the sacrifice of other areas up and down stream.

While this is a sample listing, the idea is that we do not want to waste our time and energy developing recommendations that do not have a chance of being implemented in the eyes of the decision maker. Efforts should be made to seek out such information and then frame the team's recommendations around the criteria.

DEVELOPING THE RECOMMENDATIONS

Every corporation will have its own standards in how it wants recommendations to be written. It will be the RCA team's goal to abide by these internal standards while accomplishing the objectives of the RCA's Team Charter.

The core team members, at a predetermined location and time, should discuss recommendations. The entire meeting should be set aside to concentrate on recommendations alone. At this meeting the team should consider the Recommendation Acceptance Criteria (if any were obtained) and any extenuating circumstances. Remember our analogy of the detective throughout this text, always trying to build a solid case. This report and its recommendations represent our "day in court." In order to win the case, our recommendations must be solid and well thought out. But foremost, they must be accepted, implemented, and effective in order to be successful.

At this team meeting, the objective should be to gain team consensus on recommendations brought to the table. Team consensus is not team agreement. Team agreement means that everyone gets what he or she wants. Team consensus means that everyone can live with the content of the recommendations. Everyone did not get all of what they wanted, but they can live with it. Team agreement is rare.

The recommendations should be clear, concise, and understandable. Always have the objective in mind of eliminating or greatly reducing the impact of the cause when writing the recommendations. Every effort should be made to focus on the RCA. Sometimes we have a tendency to have pet projects that we attach to an RCA recommendation because it might have a better chance of being accepted. We liken this to riders on bills reviewed in Congress. They tend to bog down a good bill and threaten its passage in the long run. At the first sight of unnecessary recommendations, decision makers will begin to question the credibility of the entire RCA. When writing recommendations, stick to the issues at hand and focus on eliminating the risk of recurrence.

When the team develops recommendations, it is a good idea to present decision makers with multiple alternatives. Sometimes when we develop recommendations, they might be perceived as not meeting the predefined criteria given by management. If this is the case, then efforts should be made to have an alternative recommendation—a recommendation that clearly fits within the defined criteria. One thing that we never want to happen is for an issue under which the presenters have some control to stall the management presentation. Absence of an acceptable recommendation is one such obstacle, and every effort should be made to gain closure of the RCA recommendations at this meeting.

DEVELOPING THE REPORT

The report represents the documentation of the "solid case" for court or, in our circumstances, the final management meeting. This should serve as a living document in that its greatest benefit will be that others learn from it so as to avoid the recurrence of similar events at other sites within the company or organization. To this end, the professionalism of the report should suit the nature of the event being analyzed. We like to use the adage, "If the event costs the corporation $5, then perform a $5 RCA. If it costs the organization $1,000,000, then perform a $1,000,000 type of RCA."

We should keep in mind that if RCAs are not prevalent in an organization, then the first RCA report usually sets the standard. We should be cognizant of this and take it into consideration when developing our reports. Let's assume at this point that we have analyzed a "Significant Few" event and it is costly to the organization. Our report will reflect that level or degree of importance.

The following table of contents will be our guide for the report:

1. The Executive Summary
   a. The Event Summary
   b. The Event Mechanism

    c. The PROACT Description
    d. The Root Cause Action Matrix
2. The Technical Section
    a. The Identified Root Cause
    b. The Type of Root Cause
    c. The Responsibility of Executing Recommendation
    d. The Estimated Completion Date
    e. The Detailed Plan to Execute Recommendation
3. Appendices
    a. Recognition of All Participants
    b. The 5P's Data Collection Strategies
    c. The Team Charter
    d. The Team Critical Success Factors
    e. The Logic Tree
    f. The Verification Logs
    g. The Recommendation Acceptance Criteria (if applicable)

Now let's review the significance and contribution of each element to the entire report and the overall RCA objectives.

The Executive Summary

The Executive Summary is just that—a summary. It has been our experience that the typical decision makers at the upper levels of management are not nearly as concerned with the details of the RCA as they are with the results and credibility of the RCA. This section should serve as a synopsis of the entire RCA—a quick overview. This section is meant for managers and executives to review the analyzed event, the reason it occurred, what the team recommends to make sure it never happens again, and how much it will cost.

The Event Summary

The Event Summary is a description of what was observed from the point in time that the event occurred until the point in time that the event was isolated or contained. This can generally be thought of as a timeline description.

The Event Mechanism

The Event Mechanism is a description of the findings of the RCA. It is a summary of the errors that lead up to the point in time of the event occurrence. This is meant to give management a quick understanding of the chain of errors that were found to have caused the event in question.

The PROACT Description

The PROACT Description is a basic description of the PROACT process for management. Sometimes management may not be aware of a formalized RCA process being used in the field. A basic description of such a disciplined and formal process

TABLE 11.1
Sample Root Cause Action Matrix

Cause	Type	Recommendation	Responsibility	Implementation Date	Completion Date
Outdated Start-Up Procedure	Latent	Assemble team of seasoned operators to develop the initial draft of a current start-up procedure that is appropriate for the current operation.	RJL	10/20/99	11/15/99

generally adds credibility to the analysis and assures management that it was a professional effort.

The Root Cause Action Matrix

The Root Cause Action Matrix is a table outlining the results of the entire analysis. This table is a summation of identified causes, overview of proposed recommendations, person responsible for executing recommendations, and estimated completion date. Table 11.1 shows a sample Root Cause Action Matrix.

The Technical Section

The Technical Section is where the details of all recommendations are located. This is where the technical staff may want to review the details of the analysis recommendations.

The Identified Root Cause(s)

The Identified Root Cause(s) will be delineated in this section as separate line items. All causes identified in the RCA that require countermeasures will be listed here.

The Type of Root Cause(s)

The Type of Root Cause(s) will be listed here to indicate their nature as being Physical, Human, or Latent Root Cause(s). It is important to note that only in cases of intent with malice should any indications be made as to identifying any individual or group. Even in such rare cases, it may not be prudent to specifically identify a person or group in the report because of liability concerns. Normally, no recommendations are required or necessary where a Human Root is identified. This is because if we address the Latent Root or the decision-making basis that led to the occurrence of the event, then we should subsequently change the behavior of the individual. For instance, if we have identified a Human Root as "Misalignment" of a shaft (no name necessary), then the actions to correct that situation might be to provide the individual the training and tools to align the shaft properly in the future.

This countermeasure will address the concerns of the Human Root without making a specific Human Root recommendation or potentially giving the perception of blaming individuals or groups.

The Responsibility of Executing the Recommendation

This will also be listed so as to identify an individual or group that will be accountable for the successful implementation of the recommendation.

The Estimated Completion Date

The Estimated Completion Date will be listed to provide an estimated timeline for when each countermeasure will be completed, thus setting the anticipated timeline of returns on investment.

The Detailed Plan to Execute Recommendation

This section is generally viewed as an expansion of the Root Cause Action Matrix described previously. Here is where all the economic justifications, the plans to resource the project (if required), the funding allocations, etc., are located.

Appendices

Recognition of All Participants

Recognizing all participants is extremely important if our intent is to have team members participate on RCA teams in the future. It is suggested to note every person that inputs any information into the analysis in this section. All people tend to crave recognition for their successes.

The 5P's Data Collection Strategies

These strategies should be placed as an addendum or appendix item to show the structured efforts to gain access to the necessary data to make the RCA successful.

The Team Charter

The Team Charter should also be placed in the report to show that the team displayed structure and focus with regard to their efforts.

The Team Critical Success Factors

Including the Team Critical Success Factors shows that the teams had guiding principles and defined the parameters of success.

The Logic Tree

The Logic Tree is a necessary component of the report for obvious reasons. The Logic Tree will serve as a dynamic expert system (or troubleshooting flow diagram) for future analysts. This type of information will optimize the effectiveness of any corporate RCA effort by conveying valuable information to other sites with similar events.

The Verification Logs

The Verification Logs are the spine of the logic tree and a vital part of the report. This section will house all of the supporting documentation for hypothesis validation.

The Recommendation Acceptance Criteria (if Applicable)

The Recommendation Acceptance Criteria should be listed to show that the recommendations were developed around documented criteria. This will be helpful in explaining why certain countermeasures were chosen over others.

Report Use, Distribution, and Access

The report will serve as a living document. If a corporation wishes to optimize the value of its intellectual capital using RCA, then the issuing of a formal professional report to other relevant parties is absolutely necessary. Serious consideration should be given to RCA report distributions. Analysts should review their findings and recommendations and identify others in their organization who may have similar operations and therefore similar problems. These identified individuals or groups should be put on a distribution list for the report so that they are aware that this particular event has been successfully analyzed and recommendations have been identified to eliminate the risk of recurrence. This optimizes the use of the information derived from the RCA.

In our information era, instant access to such documents is a must. Most corporations have their own internal intranets. This provides an opportunity for the corporation to store these newly developed "dynamic expert systems" in an electronic format allowing instant access. Corporations should explore the feasibility of adding such information to their intranets and allowing all sites to access the information. We will discuss automating the RCA process in Chapter 13. Using RCA software like PROACT will make RCA information more accessible to stakeholders.

Whether the information is in paper or electronic format, the ability to produce RCA documentation quickly could help some organizations from a legal standpoint. Whether it is a government regulatory agency, corporate lawyers, or insurance representatives, demonstrating that a disciplined RCA method was used to identify root causes can prevent some legal actions against the corporations as well as fines from being imposed due to noncompliance with regulations. Most regulatory agencies that require a form of RCA to be performed by organizations do not delineate the RCA method to be used, but rather ensure that one can be demonstrated upon audit.

THE FINAL PRESENTATION

This is the Principal Analyst's "final day in court." It is what the entire body of RCA work is all about. Throughout the entire analysis, the team should be focused on this meeting. We have used the analogy of the detective throughout this text. In the chapter "Preserving Failure Data" (Chapter 8) we described why a detective goes to the lengths that he or she does in order to collect, analyze, and document data. Our conclusion was that the detective knows he or she is going to court and that the lawyers must present a solid case in order to obtain a conviction.

Our situation is not much different. Our court is a final management review group that will decide if our case is solid enough to approve the requested monies for implementing recommendations.

Realizing the importance of this meeting, we should prepare accordingly. Preparation involves the following steps:

1. Have the professionally prepared reports ready and accessible.
2. Strategize for the meeting by knowing your audience.
3. Have an agenda for the meeting.
4. Develop a clear and concise professional presentation.
5. Coordinate the media to use in the presentation.
6. Conduct "dry runs" of the final presentation.
7. Quantify the effectiveness of the meeting.
8. Prioritize recommendations based on impact and effort.
9. Determine "next step" strategy.

We will address each of these individually and in some depth to maximize the effectiveness of the presentation and ensure that we get what we want.

Have the Professionally Prepared Reports Ready and Accessible

At this stage the reports should be ready, in full color and bound. Have a report for each member of the review team as well as for each team member. Part of the report includes the logic tree development. The logic tree is the focal point of the entire RCA effort and should be graphically represented as such. The logic tree should be printed on blueprint-sized paper, in full color and laminated if possible. The logic tree should be proudly displayed on the wall in full view of the review committee. Keep in mind that this logic tree will likely serve as a source of pride for the management to show other divisions, departments, and corporations how progressive their area is in conducting RCA. It will truly serve as a trophy for the organization.

Strategize for the Meeting by Knowing Your Audience

This is an integral step in determining the success of the RCA effort. Many people believe that they can develop a top-notch presentation that will suit all audiences. This has not been our experience. All audiences are different and therefore have different expectations and needs.

Consider our courtroom scenario again. Lawyers are courtroom strategists. They will base their case on the make-up of the jury and the judge presiding. When the jury has been selected, the lawyers will determine their backgrounds, whether they are middle-class or upper class, etc. What is the ratio of men to women? What is the ethnic make-up of the jury? What is the judge's track record on cases similar to this one? What were the bases of the previous cases on which the judge has ruled? Apply this to our scenario and we begin to understand that learning about the people we must influence is a must.

In preparing for the final presentation, determine which attendees will be present. Then learn about their backgrounds. Are they technical people, financial people, or perhaps marketing and sales people? This will be of great value because making a technical presentation to a financial group would risk the success of the meeting.

Next we must determine what makes these people "tick." How are these people's incentives paid? Is it based on throughput, cost reduction, profitability, various ratios, or safety records? This becomes very important because when making our presentation, we must present the benefits of implementing the recommendations in units that appeal to the audience. For example, "If we are able to correct this startup procedure and provide the operators the appropriate training, based on past history, we will be able to increase throughput by 1%, which will equate to $5 million."

Have an Agenda for the Meeting

No matter what type of presentation media you have, always have an agenda prepared for such a formal presentation. Management typically expects this formality and it also shows organizational skills on the part of the team. Table 11.2 is a sample agenda that we typically follow in our RCA presentations. Always follow the agenda; only divert when requested by the management team.

Notice that the last item on the agenda is "Commitment to Action." This is a very important agenda item as sometimes we tend to leave such meetings with a feeling of emptiness and we turn to our partner and ask, "How do you think it went?" Until this point we have done a great deal of work and we should not have to "wonder how it went." It is not impolite or too forward to ask at the conclusion of the meeting, "Where do we go from here?" Even a decision to do nothing is a decision and you know where you stand. Never leave the final meeting wondering how it went.

TABLE 11.2
Sample Final Presentation Agenda

#	Agenda Topic	Speaker
1	Review of PROACT Process	RJL
2	Summary of Undesirable Event	RJL
3	Description of Error Chain Found	KCL
4	Logic Tree Review	KCL
5	Root Cause Action Matrix Review	WTB
6	Recognition of Participants Involved	WCW
7	Question and Answer Session	ALL
8	Commitment to Action/Plan Development	RJL

Develop a Clear and Concise Professional Presentation

Research shows that the average attention span of individuals in managerial positions is about 20 to 30 minutes. The presentation portion of the meeting should be designed to accommodate this time frame. We recommend that the entire meeting last no more than 1 hour. The remaining time will be left to review recommendations and develop action plans.

The presentation should be molded around the agenda we developed earlier. Typical presentation software such as Microsoft's PowerPoint®* provides excellent graphic capabilities and also easily allows the integration of words, digital images, and animation. Remember that this is our chance to communicate our findings and recommendations. Therefore, we must be as professional as possible to get our ideas approved for implementation. The use of various forms of media during a presentation provides an interesting forum for the audience and also aids in retention of the information by the audience.

There is a complete psychology behind how the human mind tends to react to various colors. This type of research should be considered during presentation development. The use of laptops, LCD projectors, and easel pads will help in providing an array of different media to enhance the presentation. Props such as failed parts or pictures from the scene can be used to pass around to the audience and enhance interest and retention. All of this increases the chances of acceptance of the recommendations.

Always dress the part for the presentation. Our rule of thumb has always been to dress one level above the audience. We do not want to appear too informal, but we also do not want to appear too overdressed. The key is to make sure your appearance is professional. Remember, we perform a $5 failure analysis on a $5 failure. This presentation is intended for a "Significant Few" item and the associated preparation should reflect its importance.

Coordinate the Media to Use in the Presentation

As discussed earlier, many forms of media should be used to make the presentation. To that end, coordination of the use of these items should be worked out ahead of time to ensure proper "flow" of the presentation. This is very important as lack of such preparation could affect the results of the meeting and give the presentation a disconnected or unorganized appearance.

Assignment of tasks should be made prior to the final presentation. Such assignments may include one person to manipulate the computer while another presents, a person to hand out materials or props at the speaker's request, and a person who will provide verification data at the request of management. Such preparation and organization really shines during a presentation and it is apparent to the audience.

It is also very important to understand the layout logistics of the room in which you are presenting. Nothing is worse than showing up at a conference room and

---

* PowerPoint® is a registered trademark of Microsoft.

realizing that your laptop does not work with the LCD projector. Then you spend valuable time fidgeting with it and trying to make it work in front of your audience. Some things to keep in mind to this end include the following:

1. Know how many will be in your audience and where they will be sitting.
2. Use name cards if you wish to place certain people in certain positions in the audience.
3. Ensure that everyone can see your presentation from where they are sitting.
4. Ensure that you have enough handout material (if applicable).
5. Ensure that your A/V equipment is fully functional prior to the meeting.

Like everything else about RCA, we must be proactive in our preparation for our final presentation. After all, if we do not do well in this presentation, our RCA will not be successful and thus we will not have improved the bottom line.

Conduct "Dry Runs" of the Final Presentation

The final presentation should not be the testing grounds for the presentation. No matter how prepared we are, we must display some modesty and realize that there is a possibility that we may have holes in our presentation and our logic.

We are advocates of at least two "dry runs" of the presentation being conducted prior to the final one. We also suggest that such dry runs be presented in front of the best and most constructive critics in your organization. Such people will be happy to identify logic holes, thereby strengthening the logic of the tree. The time to find gaps in logic is prior to the final presentation, not during. Logic holes that are found during the final presentation will ultimately damage the credibility of the entire logic tree. This is a key step in preparation for the final presentation.

Quantify the Effectiveness of the Meeting

Earlier we discussed obtaining the recommendation acceptance criteria from management prior to developing recommendations. If these criteria are provided, this offers a basis for quantifying our meeting results if our management is progressive enough to utilize quantification tools.

We recommend the use of an evaluation tool during the presentation of the recommendations that would require the management review group to evaluate each recommendation against its compliance with the predetermined recommendation acceptance criteria. Table 11.3 is a sample cross section of such an evaluation tool.

If utilized, this form should be developed prior to the final meeting. Make as many copies as there are evaluators. As shown, the recommendations should be listed on the rows and the Recommendation Acceptance Criteria should be listed across the columns. As we are making our presentation with regard to various recommendations, we will ask the evaluators to rate the recommendation against the criteria using a scale of 0 to 5. A "5" rating would indicate that the recommendation is on target and meets the criteria given by the management. A "0", on the other

TABLE 11.3
Sample Quantitative Evaluation Form

Recommendation	Must Eliminate Cause	Must Provide a 20% ROI	Must Not Interfere with Any Capital Projects on Books	Average

hand, would indicate that the recommendation absolutely does not comply with the criteria set forth.

Based on the number of evaluators, we would take averages for how each recommendation fared against each criteria item and then take the average of those items for each recommendation and obtain a total average for how well each recommendation matched all criteria. Table 11.4 is a sample of a completed evaluation form.

Once this form has been completed, it can be applied to a predetermined scale such as the one in Table 11.5.

Once this process has been completed, we will understand what corrective action will be taken, which recommendations need modification, and which were rejected. This process allows interaction with the management during the presentation. It also allows for discussions that may arise when one manager rates a recommendation against a criteria with a "0" and another rates the same with a "5". Such disparities

TABLE 11.4
Sample Completed Quantitative Evaluation Form

Recommendation	Must Eliminate Cause	Must Provide a 20% ROI	Must Not Interfere with Any Capital Projects on Books	Average
Modify maintenance procedures to enhance precision work.	3.5	4	5	4.17
Design, implement, and instruct lubricators on how, when, and where to lubricate.	3	5	2	3.33
Develop a 3-hour training program to educate lubricators on the science of tribology.	4	5	4	4.33

TABLE 11.5
Sample Evaluation Scale

Average Score	Accept as Is	Accept with Modification	Reject
≥ 3.75	X		
≥ 2.5 < 3.75		X	
< 2.5			X

beg an explanation as to why the perspectives are so far apart. This is an unbiased and nonthreatening approach to quantifiably evaluating recommendations in the final presentation. It has been our experience, though, that only a very open-minded management would participate in such an activity.

Prioritize Recommendations Based on Impact and Effort

Part of getting what we want from such a presentation involves presenting the information in a "digestible" format. For instance, if you have completed an RCA and have developed 59 recommendations, the next task is to get them completed. As we well know, if we put 59 recommendations on someone's desk, there is a reduced likelihood that any will get done. Therefore, we must present them in a "digestible" manner. We must present them in such a format that it does not appear to be as much as it really is. How do we accomplish this task?

We utilize what we call an impact-effort priority matrix. This is a simple three-by-three table with the X-axis indicating impact and the Y-axis indicating effort to complete. Figure 11.1 is as example of such a table.

Let's return to our previous scenario of having 59 recommendations. At this point we can separate the recommendations over which we have direct control to

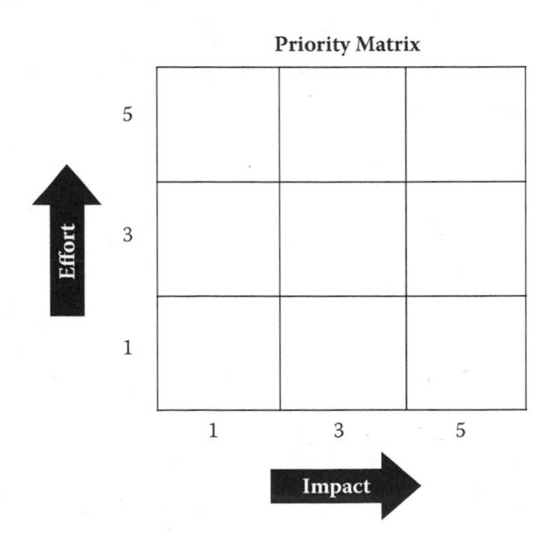

FIGURE 11.1 Impact-Effort Priority Matrix.

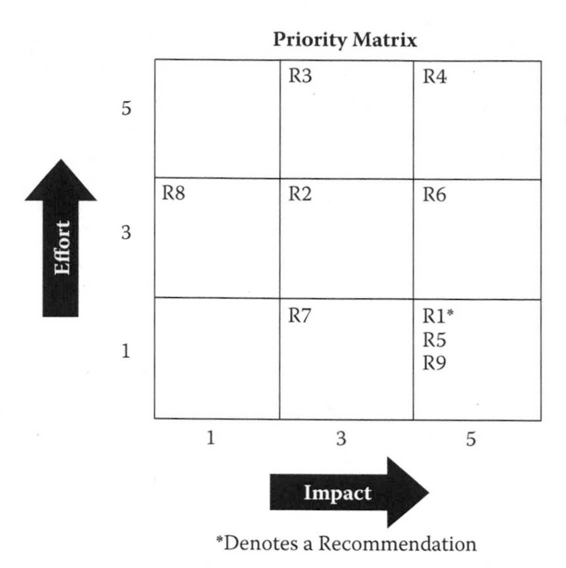

FIGURE 11.2 Completed Impact-Effort Priority Matrix.

execute and determine them to be high impact, low effort recommendations. Maybe we deem several other recommendations as requiring the approval of other departments; therefore, they may be a little more difficult to implement. Finally, maybe we determine that some recommendations require that a shutdown occur before the corrective action can be taken. Therefore, these recommendations are more difficult to implement. This is a subjective evaluation that breaks down the perception of too many recommendations into manageable and accomplishable tasks. A completed matrix may look like Figure 11.2.

Determine Next Step Strategy

The ultimate result we are looking for from this step (Communicate Findings and Recommendations) is a corrective action plan. This entire chapter dealt with selling the recommendations and gaining approvals to implement them. After the meeting we should have recommendations that have been approved, individuals assigned to execute them, and timelines in which to have them completed. The next phase we will explore is the effectiveness of the implementation and overall impact on bottom-line performance.

12 Tracking for Bottom-Line Results

Consider what we have accomplished thus far in the PROACT RCA process:

1. Established management systems to support RCA
2. Conducted an Opportunity Analysis (OA)
3. Developed a data preservation strategy
4. Organized an ideal RCA team
5. Utilized a disciplined method to draw accurate root causes
6. Prepared a formal RCA report and presentation for management
7. Defined which corrective actions will be implemented

This is an immense amount of work and an accomplishment in and of itself. However, success is not defined as identifying root causes and developing recommendations. Something has to improve as a result of implementing the recommendations!

Always keep in the back of our minds that we are continually selling our need to survive, whether it is in society or in our organization. We must be constantly proving why we are more valuable to the facility than others. Tracking for results actually becomes the measurement of our success in the RCA effort. Therefore, since this is a reflection of our work, we should be diligent in measuring our progress because it will be viewed as a report card of sorts. Once we establish successes, we must exploit them by publicizing them for maximum personal and organizational benefit. The more people who are aware and recognize the success of our efforts, the more they will view us as people to depend on in order to eliminate problems. This makes us a valuable resource to the organization. Make note that if we are successful at RCA, the reward should be that we get to do it again. This will result in the various departments or areas requesting the RCA service from us. While this is a good indication, there can be drawbacks.

For instance, we have been trained to work on the "Significant Few" events that are costing 80% of the organization's losses. Under the described circumstances, we may have numerous people asking us to solve their smaller problems, which are not necessarily important to the organization as a whole. Therefore, when we decline, we may be viewed as not being a team player because we insisted on sticking to the "Significant Few" list from the Opportunity Analysis. These are legitimate concerns that we should address with our Champions and Drivers.

Let's pick up from the point where management has approved various recommendations of ours in our final meeting. Now what happens? We must consider each of the following steps:

1. Getting proactive work orders accomplished in a reactive environment
2. Sliding the proactive work scale
3. Developing tracking metric(s)
4. Exploiting successes
5. Creating a critical mass
6. Recognizing the lifecycle effects of RCA on the organization

GETTING PROACTIVE WORK ORDERS ACCOMPLISHED IN A REACTIVE ENVIRONMENT

Unless approved recommendations are implemented, we certainly cannot expect phenomenal results. Therefore, we must be diligent in our efforts to push the approved recommendations all the way through the system. One roadblock that we have repeatedly run into is the fact that people generally perceive recommendations from RCAs as improvement work or proactive work. In the midst of a reactive backlog of work orders, a proactive one does not stand a chance for implementation.

Most Computerized Maintenance Management Systems (CMMS), or their industry equivalents, possess a feature by which work orders are prioritized. Naturally, anyone who creates a work request thinks that their work is more important than anyone else's; therefore, they put the highest priority on the work request. The priority ranking system of any work order system goes something like this (or equivalent):

E = Emergency—Respond Immediately
1 = 24-Hour Response Required
2 = 48-Hour Response Required
3 = 1-Week Response Required

What normally happens with such prioritization systems is that a large number of corrective work requests are entered as "E" or emergency events requiring the original schedule to be broken in order to accommodate them. Usually the preventive and predictive inspections are the first items to get removed from the schedule—the proactive work!

Given this scenario, what priority would a recommendation from an RCA have? Typically a "4"! Such work is deemed as backburner work and it can wait because the event is not occurring now. This is an endless cycle if the chain is not broken. This is like waiting to fix the hole in the roof until it rains.

We mentioned earlier that management systems must be put into place to support RCA efforts. This is one system that must be in place prior to even beginning RCA. If the recommendations are never going to be executed, then the RCA should never begin. Accommodations must be made in the work order system to give proactive work a fair chance of being accomplished against the reactive work. This will involve planners and schedulers to agree that a certain percentage of the maintenance resources must be allocated to executing the proactive work, no matter what. This is hard to do, both in theory and practice. But the fact of the matter is that if we do not take measures to prevent the recurrence of undesirable events, we are acknowledging defeat against them and accepting reaction as the maintenance strategy. If we do not initially allocate some degree of resources to proactive work, we will always be stuck in a reactive cycle.

The answer to the above paradox can be quite simple. We have seen companies simply identify a designation for proactive work and ensure that the planners and schedulers treat them as if they are an "E" ticket with the resources they have set aside to address such opportunities. Maybe it's a "P" for proactive work or a block of worker numbers. Whatever the case may be, consideration must be given to making sure that proactive work orders generated from RCAs are implemented in the field.

This needs to be a priority for the company and tracked as such. We need to track the amount of proactive work being done on a monthly basis. If the level of proactive work is insufficient, we need to make our plant Driver and Champion aware so they can address the issues. Most organizations do not like change. We are all in favor of improving things as long as we do not have to "change." Utilizing metrics to measure our level of proactive work will demonstrate how committed we are to improvement and defect elimination.

SLIDING THE PROACTIVE WORK SCALE

As we hear all the time, the most common objection to performing RCA in the field is that we do not have the time. When we look at this objection introspectively, we find that we do not have the time because we are too busy reacting to failures and repairs. This truly is an oxymoron. RCA is designed to eliminate the need to react to unexpected failures. Managements must realize this and include RCA as part of the overall plant strategy.

One way we have seen this done is through an interactive board game developed originally within DuPont®* and now licensed through a company in Kingwood, TX, called The Manufacturing Game®.† Organizational development experts within DuPont® developed this game. It is an innovative way to involve all perspectives of a manufacturing plant. When we played The Manufacturing Game® we found it to be an invaluable tool for demonstrating why a facility must allocate some initial resources to proactive work in order to remain competitive and in business. The Manufacturing Game® demonstrates why proactive activities are needed to eliminate the need to do work and RCA expresses how to actually do it.

Proaction and reaction should be inversely proportional. The more proactive tasks that are performed the less reactive work there should be. Therefore, all the personnel we currently have conducting strictly reactive work will now have more time to face the challenges of proactive work (Figure 12.1). We have yet to see a facility that admittedly has all the resources it would like to have to conduct proactive tasks such as visual inspections, predictive maintenance, preventive maintenance, RCA, lubrication, etc. It does not have these resources now because they are in reactionary situations. As the level of proaction increases, the level of reaction will decrease. This

---

* DuPont is a registered trademark of the E. I. DuPont de Nemours & Co.

† ® 1998 The Manufacturing Game.

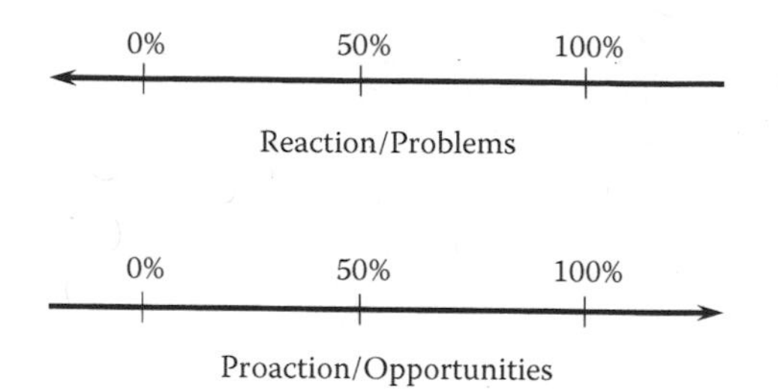

FIGURE 12.1 Inverse relationship between reaction and proaction.

is a point where we gain control of the operation and the operation does not control us! Research demonstrates that a reactive plant spends 25% more on routine maintenance than its counterparts in the proactive domain. It has also been proven that there is a direct correlation between the amount of money we spend on maintenance and the losses associated with production disruptions. Some studies suggest that for every dollar that is spent on maintenance there is a $4 to $10 loss in production. This does not even address the safety and environmental issues linked to reactive work environments.

DEVELOPING TRACKING METRICS

Recognizing the inverse relationship between proaction and reaction, we must focus on how to measure the effects of implemented recommendations. This is generally not a complex task because typically there was an existing measurement system in place that identified a deficiency in the first place. By the time the RCA is completed and the causes all identified, the metric to measure usually becomes obvious.

Let's review a few circumstances to determine appropriate metrics:

1. *Mechanical*—We experience a Mean-Time-Between-Failure (MTBF) of 3 months on a centrifugal pump. We find that various causes that include a change of service within the past year, a new bearing manufacturer is being used, and the lubrication task has been shifted to operations personnel. We take corrective actions to properly size the pump for the new service, ensure that the new bearings are appropriate for the new service, and monitor the lubrication tasks to confirm that they are being performed in a timely manner. With all these changes, we now must measure their effectiveness on the bottom line. We knew we had an undesirable situation when the MTBF was 3 months; we should now measure the MTBF over the next year. If we are successful, then we should not incur any more failures during that time period due to the causes identified in the RCA. The bottom-line effect should be that savings are realized by man-hours not expended on repairing the pump, materials not used in repairing the pump, and downtime not lost due to lack of availability of the pump (Figure 12.2).
2. *Operational*—We experience an excessive amount of rework (8%) due to production problems that result in poor quality product that cannot be sold

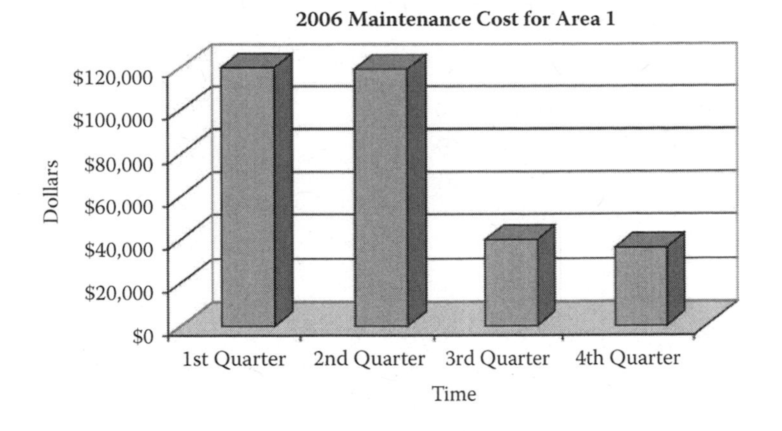

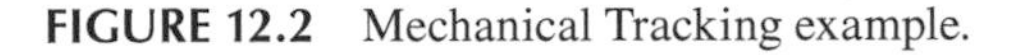
FIGURE 12.2 Mechanical Tracking example.

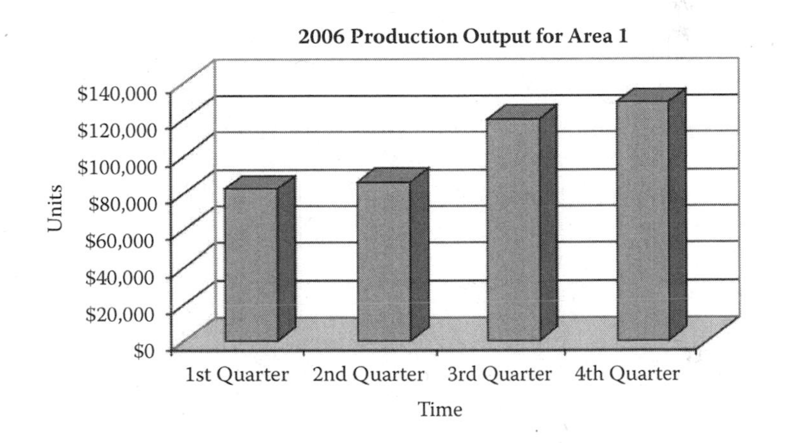

FIGURE 12.3 Operational Tracking example.

to our customers. We find as a result of our RCA that we have instrumentation in the process that is not capable of handling a recent design modification. We also find that there are inconsistencies from shift to shift in the way the same process is operated. These inconsistencies are the result of no written operating procedures. We implement the corrective action of installing instrumentation that will provide the information we require and writing a new operating procedure that insures continuity. Rework started at 8%, so after we implement our solutions we should monitor this metric and make sure it comes down significantly. The bottom-line effect is that if we are reducing rework by 8%, we should be increasing salable product by an equal amount while not incurring the costs associated with rework (Figure 12.3).

3. *Customer Service*—We experience a customer complaint rise from 2% to 5% within a 3-month period. Upon conclusion of the RCA we find that

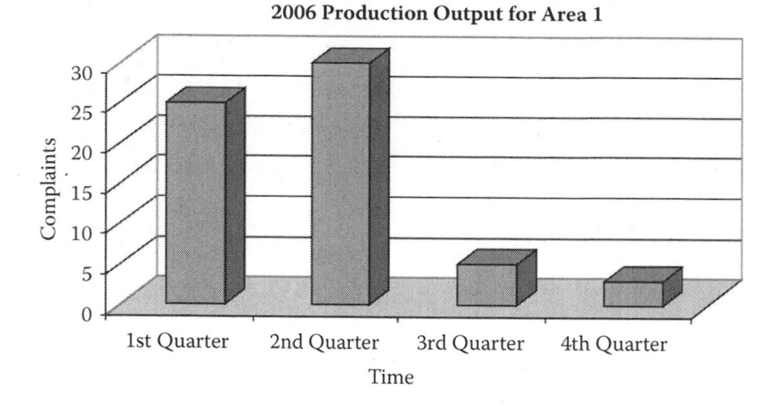

FIGURE 12.4 Customer Service Tracking example.

80% of the complaints are due to late deliveries of our product to our clients' sites. Causes are determined to be a lack of communication between purchasing and the delivery firm on pickup times and destination times. We also find that the delivery firm needs a minimum of 4-hours' notice to guarantee on-time delivery and we have been giving them only 2-hours' notice on many occasions. As a result, we have a meeting between the purchasing personnel and the dispatch personnel from the delivery firm. A mutually agreed upon procedure is developed to weed out any miscommunications. Purchasing further agrees to honor their agreement with the Delivery Company in providing a minimum 4-hour notice. Exceptions will be reviewed by the delivery firm but cannot be guaranteed. The metric we could use to measure success will be the reduction in customer complaints due to late deliveries (Figure 12.4).

4. *Safety*—We experience an unusually high number of incidents of back sprain in a package delivery hub. As a result of the RCA we find causes such as lack of training in how to properly lift using the legs, lack of warming up the muscles to be used, and heavy package trucks being assigned to those not experienced in proper lifting techniques. Corrective actions include a mandatory warm-up period prior to the shift start, attendance at a mandatory training course on how to lift properly, passing of a test to demonstrate skills learned, and modifying truck assignments to ensure that experienced and qualified loaders/unloaders are assigned to more challenging loads. Metrics to measure can include the reduction in the number of monthly back sprain claims and also the reduction in insurance costs and workman's compensation to address the claims (Figure 12.5).

The pattern of metric development described shows that the metric that initially indicated that something was wrong can also be (and usually is) the same metric that can indicate that something is improving. Sometimes this phase seems too simple and therefore it cannot be used. Then we start our "paralysis by analysis" paradigms and

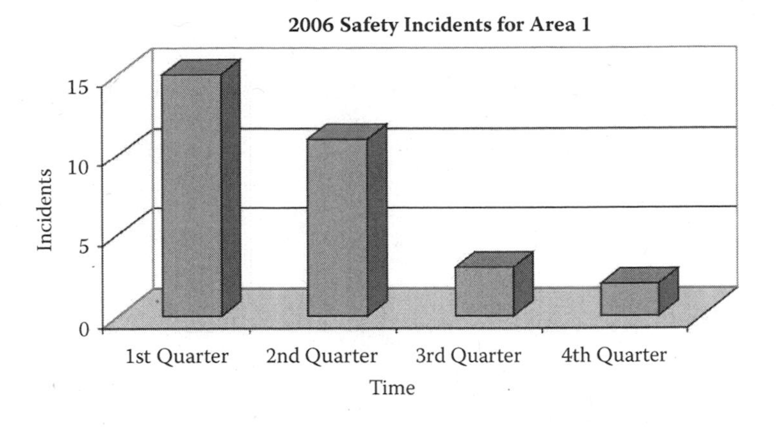

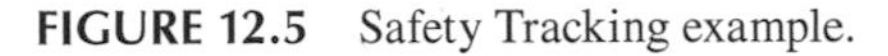
FIGURE 12.5 Safety Tracking example.

develop complex measurement techniques, which can be overkill. Not to say that they are never warranted, but we should be sure to not complicate issues unnecessarily.

EXPLOITING SUCCESSES

If no one knows the successes generated from RCA, the initiative will have a tough time moving forward and the organization will not benefit from the effects of the analyses. Like any new initiative in an organization, skepticism abounds as to its survival chances. We discussed earlier the "program-of-the-month" mentality that is likely to set in after the introduction of such initiatives. To combat this hurdle, we need to exploit successes from RCA to improve the chances that the initiative will remain viable and accepted by the work population. Without this participation and acceptance, the effort is often doomed.

How do we effectively exploit such successes? One of the main ways we do this for our clients is through high exposure mediums. High exposure mediums include such media as report distribution, internal newsletters, corporate newsletters, company intranets, presentation of success at trade conferences, written articles for trade publications, and finally exposures in texts such as this one with successes demonstrated through the use of case histories. Exploitation serves a dual purpose—it gives recognition to the corporation as a progressive entity that utilizes its workforce's brainpower, and it provides the analyst and core team recognition for a job well done. This will be the motivator for continuing to perform such work. Without recognition, we tend to move on to other things because there is no glory in this type of work.

Let's explore the different media we just mentioned.

1. *Report Distribution*—As discussed in the reporting section, to optimize the impact of RCA the results must be communicated to the people who can best use the information. In the process of doing this, we are also communicating to these facilities that we are doing some pretty good work in the name of RCA and that our people are being recognized for it.

2. *Internal Newsletters*—Most corporations have some sort of a newsletter. These newsletters serve the same purpose as a newspaper—to communicate useful information to its readers. Most publishers of internal newsletters, with whom we have never dealt, would welcome such success stories for use in their newsletters. That is what the newsletter is for; therefore, we should take advantage of the opportunity.
3. *Corporate Newsletters*—Again, most corporations we deal with have some type of corporate newsletter. It may not be published as frequently as the internal newsletter, but nonetheless, it is published. These types of newsletters focus on the "big" picture when compared to internal newsletters and may include more articles geared toward financials, overseas competition, etc. However, they too are looking for success stories that can demonstrate how to save the corporation money and recognize sites that are exemplary.
4. *Presentations at Trade Conferences*—This is a great form of recognition for both the individual (and team) and the corporation. For some analysts, this is their first appearance in a public forum. While some may be hesitant at the public speaking aspect of the event, they are generally very impressed with their ability to get through it and receive the applause of an appreciative crowd. They are also more prone to want to do it again in the future. Trade conferences thrive on the input of the companies involved in the conference. They are made up of such successes, and the conference is a forum to communicate the valuable information to others who can learn from it.
5. *Articles in Trade Publications*—As we continue along these various forms of media, the exposures become more widespread. In speaking of trade publications, we are talking about exposure to thousands of individuals in the circulation of the magazine. The reprints of these articles tend to be viewed as trophies to the analysts, who are not used to such recognition. As a matter of fact, when we have such star client analysts who have written an article of their success, we frame the reprint and send it to the analyst for display in his or her office. It is something the analyst should be proud of as an accomplishment in his or her career.
6. *Case Histories in Technical Text*—As you will read in the remainder of this text, we solicited responses from our client base on interested corporations that would like to let the general public know of the progressive work they are doing in the area of RCA and how their workforce is making an impact on the bottom line. As most any corporation will attest, no matter what the initiative is or what the new technology may be, without a complete understanding by the workforce of how to use the new information and its benefits (personally and for the corporation), it likely will not succeed. Buy-in and acceptance produce results—not intentions or expectations of the corporation.

CREATING A CRITICAL MASS

When discussing the term "critical mass" we are referring not only to RCA efforts, but the introduction of any new technology. It has been our firm's experience in training and implementing RCA efforts over the past 26 years that if we can create a critical mass of 30% of the people on board, the others will follow.

We have beat to death the "program-of-the-month" mentality, but it is a reality. Some people are leaders and others are followers. The leaders are generally the risk takers and the ones who welcome new technologies to try out. The followers are typically more conservative people who take the "let's wait and see" attitude. They believe that if this is another "program-of-the-month" they will wait it out to see if it has any staying power. These individuals are those who have been hyped up before about such new efforts, and possibly even participated in them, and then never heard any feedback about their work. They are in essence alienated with regard to "new" thinking and the seriousness of management to support it.

We believe that if we can get 30% of the trained RCA population to actually use the new skill in the field and produce bottom-line results, then RCA will become more institutionalized in the organization. If only 30% of the analysts start to show financial results, the dollars saved will be phenomenal—phenomenal enough to catch executives' eyes where they continue to support the effort with actions, not words. Once the analysts start to get recognition within the organization and corporation, others will crave similar recognition and start to participate.

We have found it unrealistic to expect that everyone we train will respond in the manner that we (and the organization) would like. It is realistic to expect a certain percentage of the population to take the new skills to heart and produce results that will encourage others to come on board.

RECOGNIZING THE LIFECYCLE EFFECTS OF RCA ON THE ORGANIZATION

RCA can play a major role in today's overall corporate strategies for growth. As we have referenced throughout this text, the goal should be the elimination of the recurrence of any undesirable outcomes that have occurred in the past. Many organizations set their sights, and thus their standards, on being the best "predictors" of such events and thus target the reduction in response time as the successful measure. While this is still a must in the interim, it should be a means to another end—the elimination of the recurrence. If we did not have undesirable outcomes, we would not have a need to become better predictors.

We have seen millions and millions of dollars spent by corporations around the world on Reliability Centered Maintenance (RCM). In its textbook implementation, RCM is ultimately geared toward helping firms determine the criticality/function of systems and equipment in their operations and then developing a specific maintenance strategy based on that information. The end result is that we have a very in-depth understanding of our operation and what could ever possibly go wrong. Most of the industrial corporations that have embraced RCM will agree that it is very expensive to implement and extremely resource intensive. However, the yields from such efforts are typically incremental in the short term.

While we have seen many organizations grasp the RCM concept, we continue to have difficulty in convincing corporations to give equal credence to an RCA or

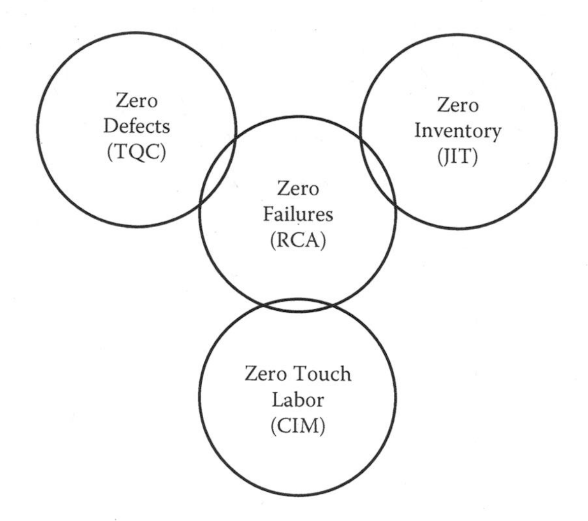

FIGURE 12.6 The Zero Imperatives.

defect elimination effort. When implemented appropriately, RCA eliminates the recurrence of events that are occurring now and that are even being compensated for in the budget. When such "chronic" issues are solved and eliminated, there is no need to budget for their occurrence anymore. The savings are off the bottom line in the same fiscal year.

We are in agreement with the concept of RCM in general. However, much time can be spent on analyzing how to combat an event that has a miniscule chance of ever occurring. RCA is geared toward working on events that have occurred and are occurring. RCM and RCA are complementary efforts toward the total elimination of undesirable events. Over the past decades we have been inundated with what we call the "Zero Imperatives" (Figure 12.6). The Zero Imperatives are the efforts associated with Zero Touch Labor, Zero Inventory, Zero Injuries, Zero Quality Defects, etc. RCA is geared toward Zero Failures or the elimination of undesirable events.

While we are all realistic about these Zero Imperatives, we realize that they are not obtainable literally, but they do provide a point toward which to strive. If our stockholders had their druthers, they would want the assets in any facility to run 24 hours a day for 365 days a year at maximum capacity in a sold-out market. This will never happen without a Zero Failure environment!

CONCLUSION

Let's face the facts—we are a human species and we are evolving. We may never be perfect, but that should not preclude us from striving to be so. We will never be error-free, but we can strive to be. Precision is a state of mind and requires the mentality to constantly strive for the next plateau.

RCA as described in this text is not a panacea. It is merely a method to assist in logical thinking to resolve undesirable events. While many of our analogies have been from the industrial world where our background lies, we hope it is clear that this RCA approach is applicable under any circumstances. Whether it is chronic or sporadic, mechanical or administrative, or in an oil refinery or a hospital, all require the same logical human thought process to resolve their respective issues.

In the following chapters we will discuss how to make this thought process more manageable. We will seek to alleviate the administrative burden of managing an RCA by providing a simple and effective software solution. While conducting RCA in a disciplined manner as we have preached in this text can be difficult, most of the time is spent sticking to the discipline and documenting the process. One of the ways we can provide an incentive to take this extra step of "discipline" is to make the task easier and more desirable. This is where the PROACT software plays its role.

Finally, we will show the "bottom-line results" achieved by those firms who had the courage to adhere to the PROACT discipline and produce phenomenal results for themselves and their companies.

13 Automating Root Cause Analysis *Introducing PROACTOnDemand℠*

PROACT* is the acronym that we have been using throughout this text to describe our Root Cause Analysis (RCA) methodology. In this chapter we will introduce the first true Internet version of our PROACT Root Cause Analysis Software Application. This means that there is no need to involve procurement and long budgetary periods because the user subscribes on an annual basis for a fee that is easily placed on a credit card. There is no installation, no capital approvals, and no annual maintenance fees. There is no need for lawyers to negotiate lengthy software licenses. There is no need to fight with IT about gaining a place on their priority list or asking for server space and an administrator because the entire program is hosted elsewhere and accessible with a browser.

In this chapter we will relate how and where there are opportunities to automate tasks that are otherwise done manually in the performance of a root cause analysis.

CUSTOMIZING PROACT FOR OUR FACILITY

One important feature about RCA software is that it should be customizable for our specific facility. This means that we would like to see accommodations for our site information (facility locations, divisions, and departments) and for our equipment listings to be input (type and class). This makes it much easier when we are working on specific analyses to be able to choose from a pick list of items that are familiar to us, and not simply pick categories that apply to any industry.

PROACT allows for such information to be input into its databases for storage and retrieval. Figure 13.1 shows the ease with which we can manually input such information if it is not readily available, or we can take existing files with such information and import it into PROACT to avoid reworking the available data.

When completed, PROACT will have stored all of the applicable facilities, their respective divisions and departments, and all the common equipment types and classes. This same process will be used to enter all of the prospective team members to participate on FMEA, OA, and RCA teams. Administrators are also able to change field labels in the event that the terminology used at the facility is different

* PROACT is a registered trademark of Reliability Center, Inc.

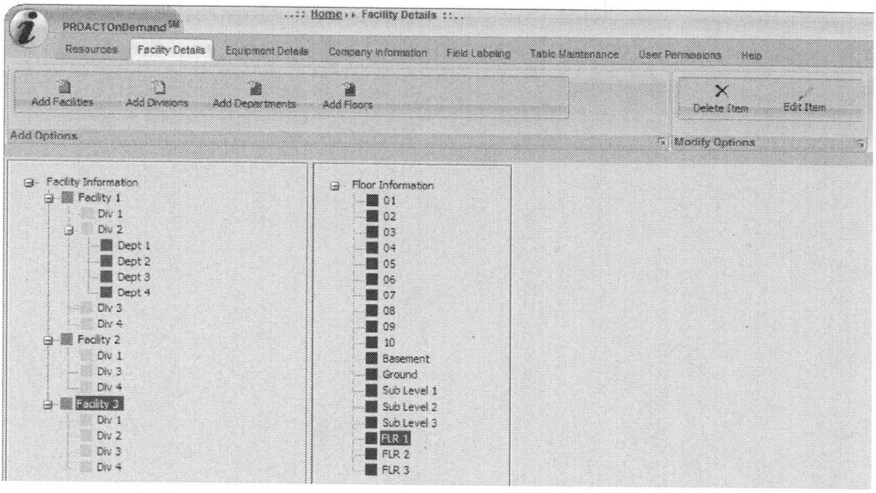

FIGURE 13.1 Facility Information screen.

(i.e., Business Units, Hospitals, etc.). A step-by-step wizard is provided to make it very easy for the user to enter this information him- or herself or request that it be imported from another system.

SETTING UP A NEW ANALYSIS IN THE NEW PROACT RCA MODULE

Once we have the administrative information stored, we can now set up a new RCA for us to start. The use of wizards in the PROACT Software is consistent for easier user orientation. Figure 13.2 shows how the user is asked to select which wizard to launch based on what the user wants to accomplish at that time. For our

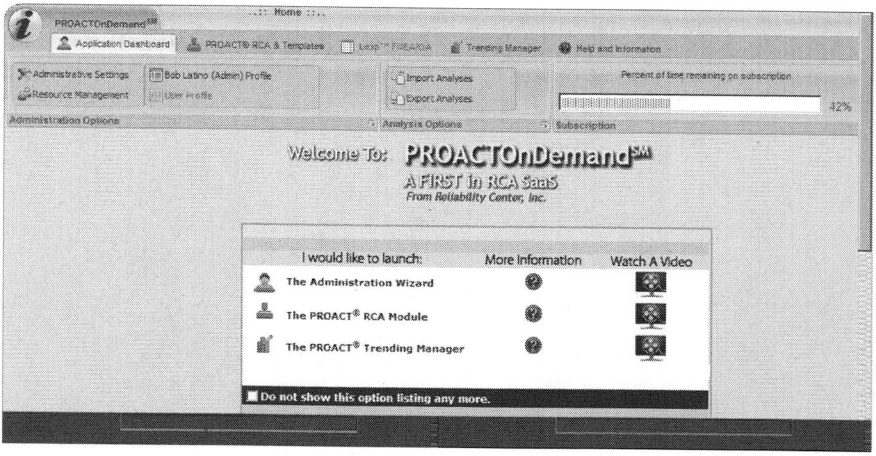

FIGURE 13.2 The PROACTOnDemand℠ Home Page.

purposes we will choose to move forward with the setting up of a PROACT Root Cause Analysis.

The New Analysis Wizard for the RCA program is a series of nine steps. Step 1 is the inputting of the new analysis name, description, and type. The type is very important. The program permits the user to pick the analysis type from a pick list, citing such choices as safety, mechanical, environmental, operational, risk, security, quality, reliability, and administrative. This will allow the eventual sorting of the analysis database on these categories if desired. Therefore, when Reliability engineers want to view all of the completed analyses on Reliability issues, they can simply sort the database on this field.

The last field on this screen deals with the Estimated Annual Cost of the Event. This is an important field if the analyst is interested in using the financial features available in the PROACT software. This will allow the program to calculate the Return on Investment (ROI) for each recommendation and also for the analysis as a whole (Figure 13.3).

Step 2 utilizes the information we input above about our facility. Step 2 involves identifying the specific location relative to the event being analyzed. This will assist us later when trying to "data mine" a database for information about the location of specific events that have been analyzed. Note that the last two fields relate to Floor, Wing, and Functional Location. Floor and Wing are more prevalent in the healthcare industry when describing locations within a hospital. Functional Location is a term often used in large Enterprise Asset Management (EAM) systems. This term refers

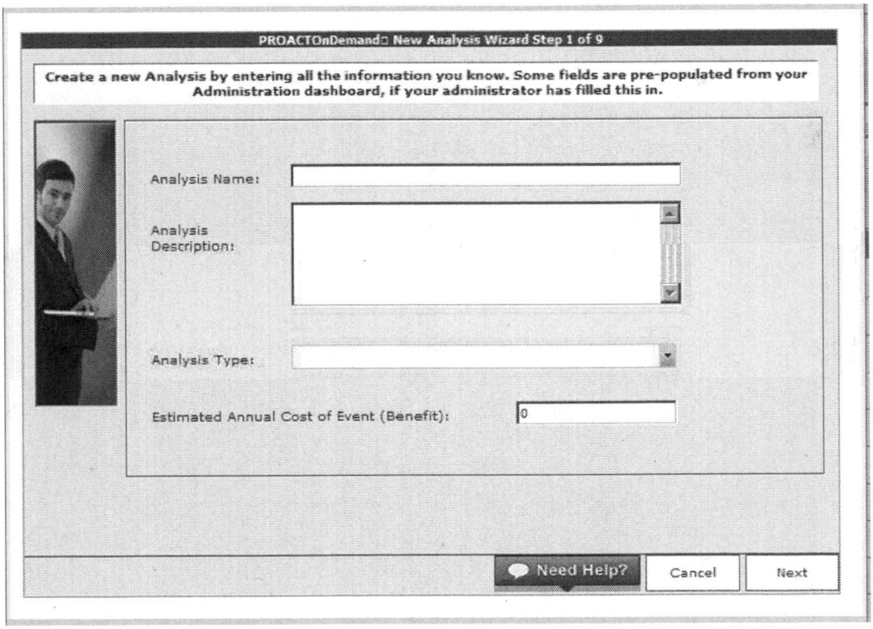

FIGURE 13.3 Setting up a new RCA, Step 1.

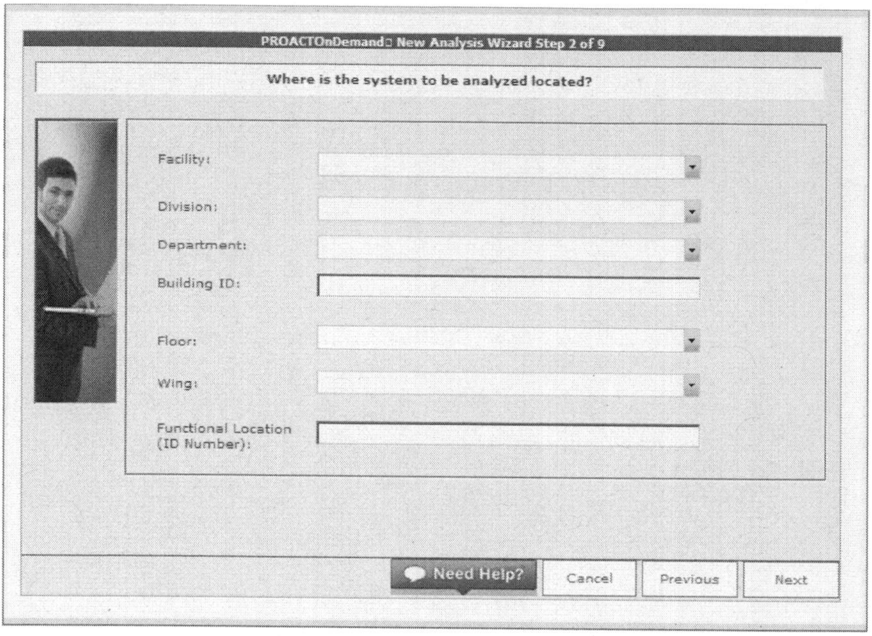

FIGURE 13.4 Setting up a new RCA, Step 2.

to a grid type of system, which indicates the location of a piece of equipment and its identification number (Figure 13.4).

Step 3 further documents the details of the event by seeking to see if there was any equipment involved. If applicable, the analyst at this point would input the pertinent equipment type, class, and manufacturer, if desired (Figure 13.5).

Step 4 is where the Principal Analyst will define the team's Critical Success Factors (CSFs). These are the seven or eight guidelines that the team agrees to abide by in order to be successful. Several default CSFs are provided, but custom ones can be entered and added to the database (Figure 13.6).

Step 5 is where the team charter is entered. This is the one-paragraph statement about why the team is together. This defines the purpose of the team. A boilerplate team charter is provided, if desired. Also, the analyst can expand on this team charter and offer extraneous information. Oftentimes this free-form space is used to describe the factors triggering the analysis to be conducted. For instance, the event may be an OSHA recordable or a Sentinel Event under The Joint Commission (TJC) guidelines (Figure 13.7).

Step 6 involves the setting up of team members for the specific analysis at hand. Again, a prepopulated database will exist with our company's personnel. We will then be able to pick and choose who would be most suitable for this analysis. Sort and filter options are available for searching the team pool listing, which in some cases can be populated with hundreds of names. Note that the person who logged into PROACT to set up the analysis will automatically be assigned as the Principal

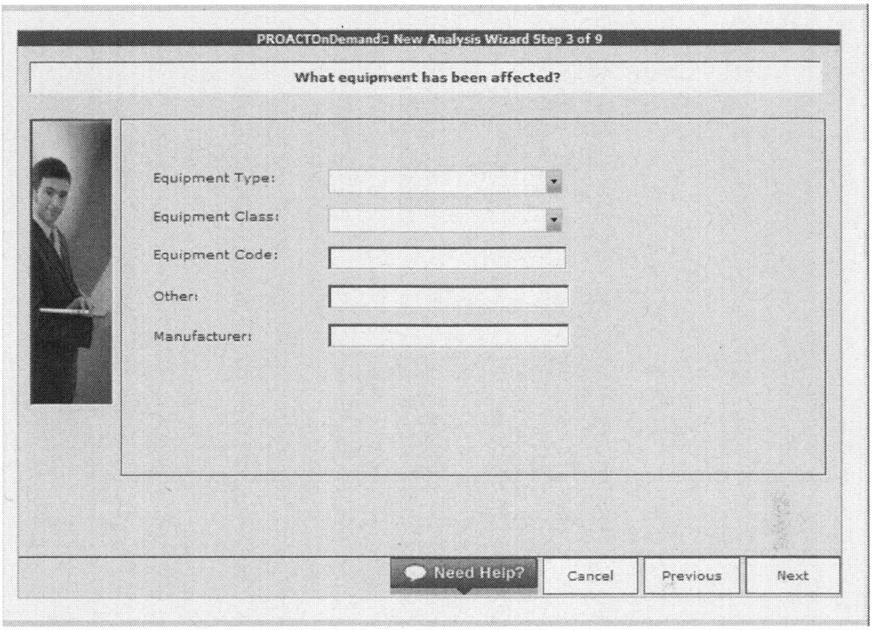

FIGURE 13.5 Setting up a new Analysis, Step 3.

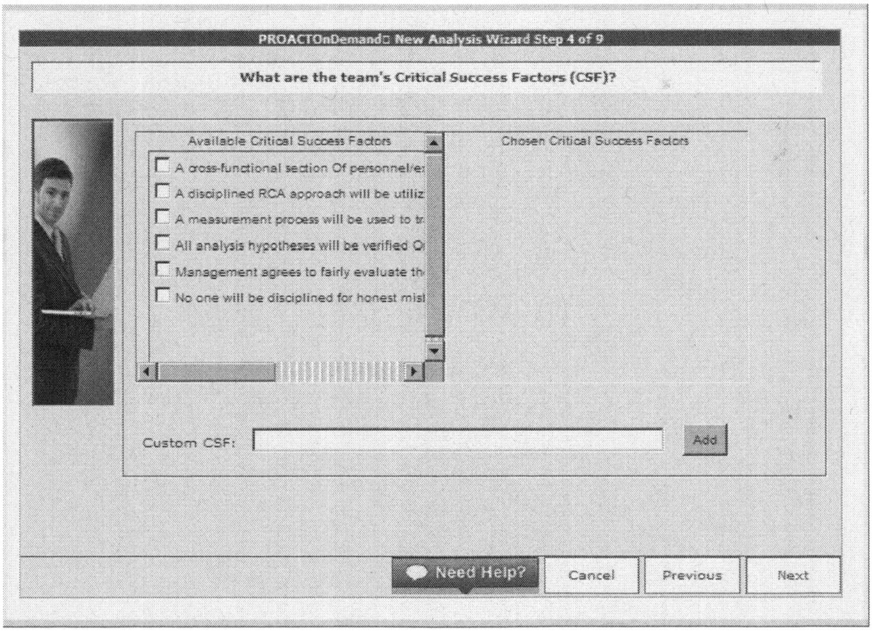

FIGURE 13.6 Setting up a new Analysis, Step 4.

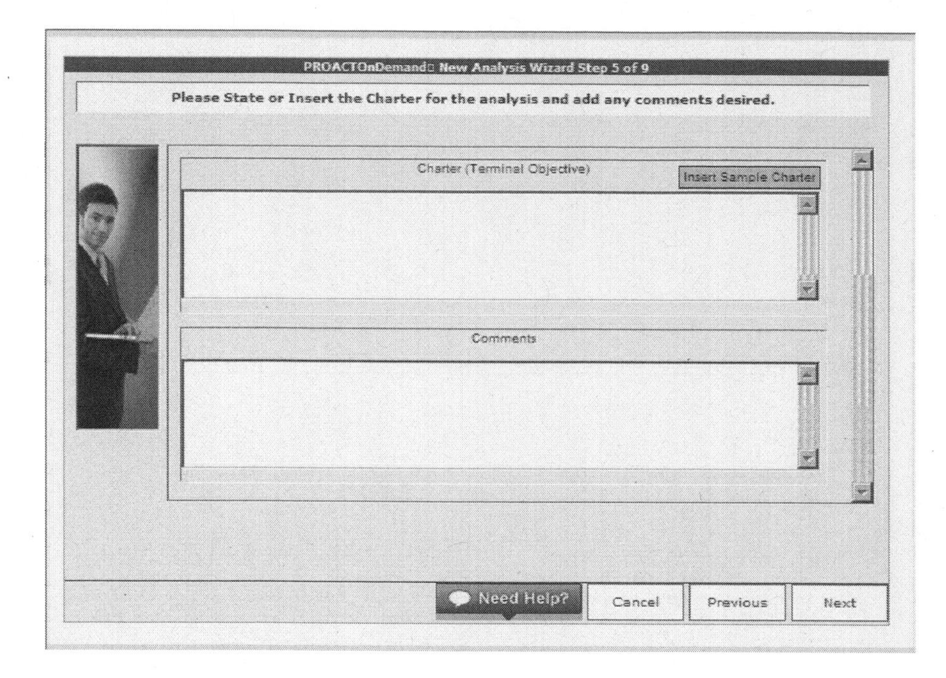

FIGURE 13.7 Sample Team Charter, Step 5.

Analyst. This can be changed down the road, but it is defaulted in the beginning (Figure 13.8).

If Team Members are selected, the Principal Analyst will have the option of assigning various permissions to each team member. For instance, a team member may be a vendor or Original Equipment Manufacturer (OEM) representative. We may not want such outsiders to be able to make changes in the analysis but to review it only, so we will give them "read only" permissions (Figure 13.9).

Step 8 determines when the analysis will start and when it will be completed. Again, a 45-day default period is built in from the *start date* and is subject to be changed by the Principal Analyst (Figure 13.10).

Step 9 is merely a confirmation that the analysis has been created successfully and that we are ready to enter into the analysis itself (Figure 13.11).

Next we will venture into the "PR" of PROACT and learn how to automate the data collection tasks.

AUTOMATING THE PRESERVATION OF EVENT DATA

In Chapter 8 we discussed the manual approach to preserving event data utilizing our 5P's Data Collection Strategy Forms. While effective, it can lack efficiency because of the organizational skills required to manage the paperwork. Also, from an efficiency standpoint, manual methods require double handling of data, which is non-value-added work. Whenever we write down information, it will eventually have to be re-entered into a computer for final presentation. Automation provides an

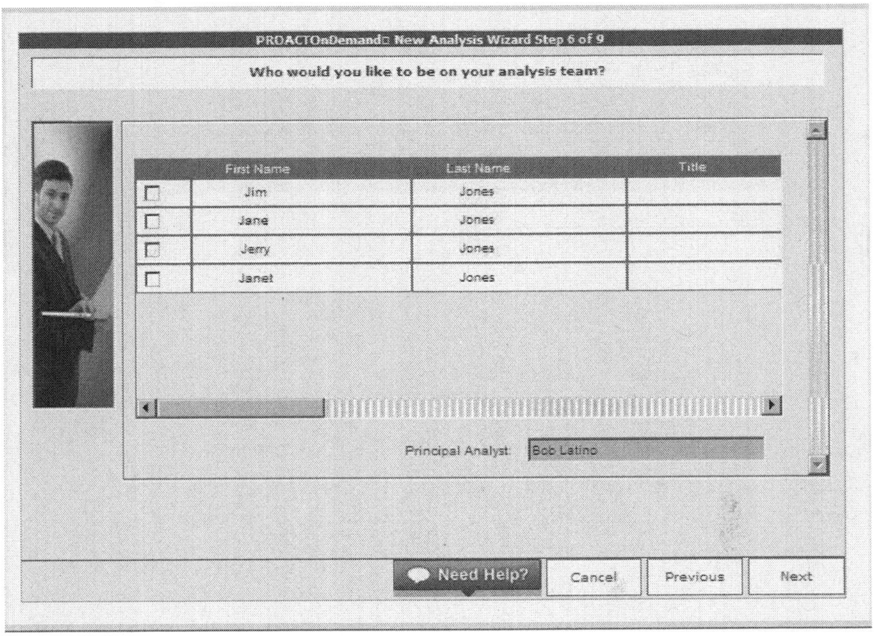

PROACTOnDemand□ New Analysis Wizard Step 6 of 9

Who would you like to be on your analysis team?

	First Name	Last Name	Title
□	Jim	Jones	
□	Jane	Jones	
□	Jerry	Jones	
□	Janet	Jones	

Principal Analyst: Bob Latino

Need Help? Cancel Previous Next

FIGURE 13.8 Setting up a new Analysis, Step 6.

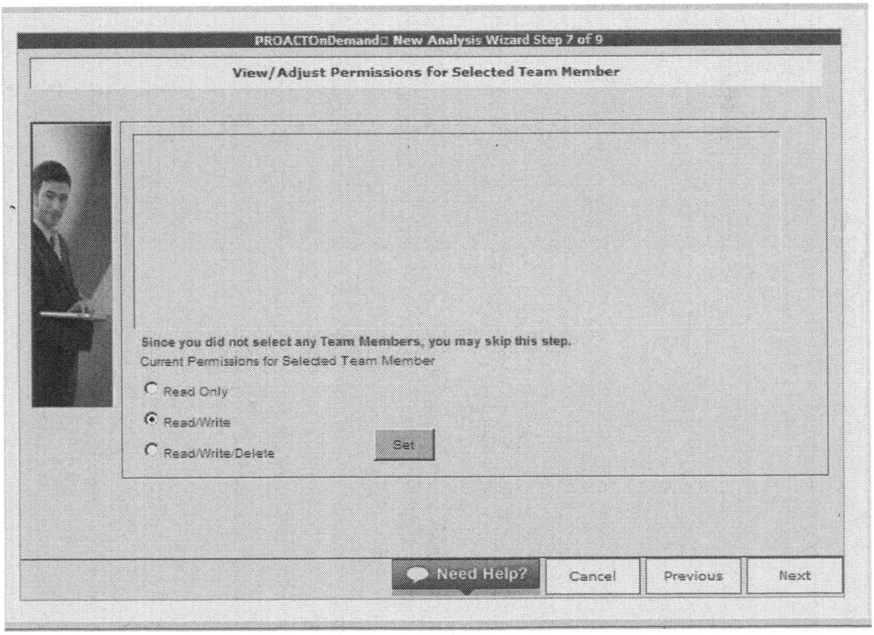

PROACTOnDemand□ New Analysis Wizard Step 7 of 9

View/Adjust Permissions for Selected Team Member

Since you did not select any Team Members, you may skip this step.

Current Permissions for Selected Team Member

Read Only

Read/Write

Read/Write/Delete

Set

Need Help? Cancel Previous Next

FIGURE 13.9 Granting team member permissions, Step 7.

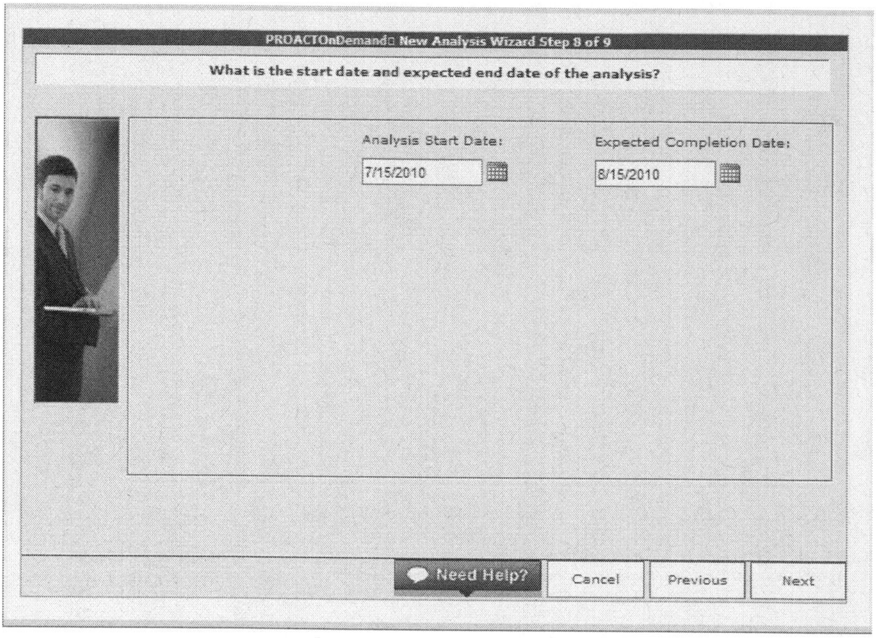

PROACTOnDemand New Analysis Wizard Step 8 of 9

What is the start date and expected end date of the analysis?

Analysis Start Date: 7/15/2010

Expected Completion Date: 8/15/2010

Need Help? Cancel Previous Next

FIGURE 13.10 Setting start and end dates, Step 8.

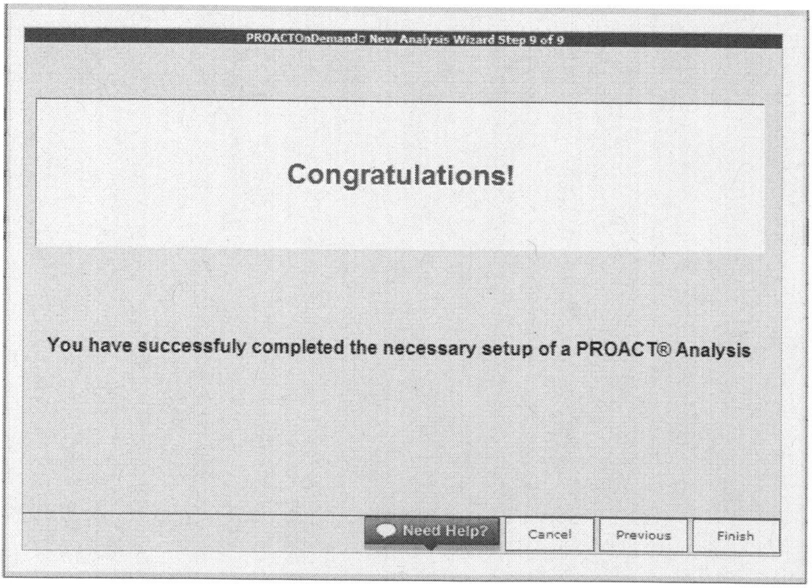

PROACTOnDemand New Analysis Wizard Step 9 of 9

Congratulations!

You have successfuly completed the necessary setup of a PROACT® Analysis

Need Help? Cancel Previous Finish

FIGURE 13.11 Positive reinforcement, Step 9.

5P's Data Collection Form

Analysis Name: ______________________________

Data Type: People, Parts, Position, Paper, Paradigms (circle one)

Champion: ______________________________

(Person that ensures all data assigned below is collected by due date)

#	Data to Be Collected	How Data Will Be Obtained (Data Collection Strategy)	Person Responsible	Date to Be Collected By

FIGURE 13.12 5P's manual data collection form.

opportunity to eliminate these inefficiencies. To refresh our memories, the manual form appears as shown in Figure 13.12.

So now that we know what information is required and the format in which we desire it, the automation requirements are determined. The screen shot in Figure 13.13 is from PROACT and shows the tabs associated with the acronym. Once the New Analysis Wizard is completed, the analyst will be defaulted into the Preserve Data Opening Screen.

PROACT is basically a glorified electronic data collector that takes out the double handling of data and streamlines the administrative tasks associated with managing an RCA. Experience demonstrates that the use of such an electronic tool to assist in RCAs will cut the analysis time in half compared with manual, paper-based approaches. In this location the analyst can add, delete, edit, or duplicate the preserve records. The analyst can also file link actual evidence to each of the records for documentation purposes. The regulatory agencies really appreciate this capability. A new feature of PROACT is the ability to automatically e-mail team members who

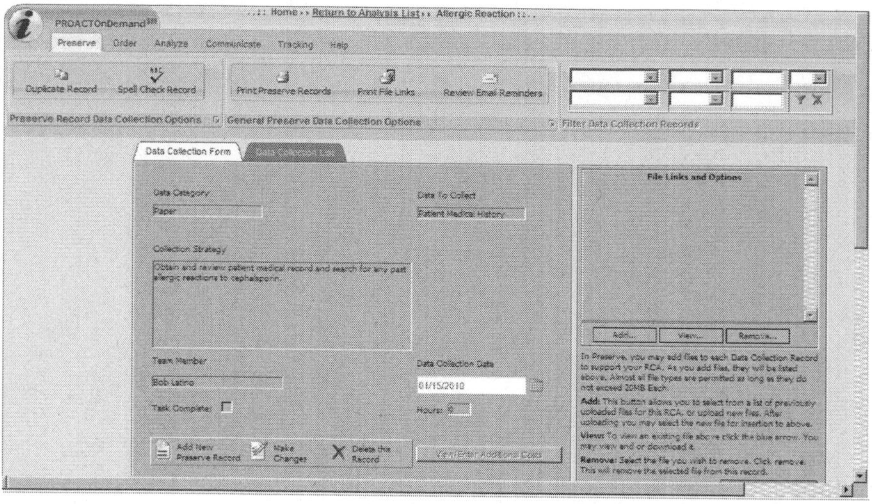

FIGURE 13.13 PROACT Analysis Introduction screen.

are assigned any task within the entire program. The Principal Analyst can set the default time periods for when they will get these notices, but this feature is a good checks-and-balances system to ensure that the analysis is progressing. Lastly, at the end of the team meeting in which team members are strategizing how to get certain data, they can immediately print a listing of what needs to be collected, by whom, and when. This saves some poor soul from having to take easel pad paper back to his or her office and enter the data into a spreadsheet (Figure 13.14).

PROACT can also track team members' time and cost to participate on the team and complete their assigned tasks. Oftentimes we must justify our RCA efforts, and having such necessary information to complete an ROI calculation can be handy in making a business case. In every location in PROACT where someone can be assigned a task, the analysts can log the time it took to complete the task and any other associated costs. PROACT will take that time and multiply it against a hidden pay scale to arrive at a total cost (Figure 13.15).

The first team meeting, as described in Chapter 9 ("Ordering the Analysis Team"), involves a brainstorming session of the core team. The team assembles and, based on the given facts at hand, starts to develop a list of data necessary to collect in order to start the analysis. This type of automation is most effective if a laptop is available in the meeting with an operator/recorder entering data as it is offered. The ideal situation is the use of an LCD projector with the laptop so the entries are seen on the screen and everyone can be assured that their information was transcribed accurately.

As the type of data to collect is entered, a team member should be assigned to obtain that information using a certain collection strategy. A time frame should be assigned to focus the team and forge a progression of the analysis.

Using an automation tool like PROACT, especially in a team format, tends to maintain interest and, more importantly, organization of the entire RCA as we go through the analysis process.

Print Preserve Information

77% | 1 of 2 | Ready | Close

Preserve Tasks for Analysis: Conveyor Roller Failures

Category	Data	Strategy	Team Member	Date	Completed	Hours
Paper	Maintenance Histories	Obtain maintenance histories for conveyor rollers on the packaging line. Collect data for past two years.	Bob Latino	5/6/2001	No	
Paper	Maintenance Procedures	Obtain maintenance procedures for how to repair and lubricate conveyor roller bearings.	David Craig	5/6/2001	No	
Paper	Purchasing Data	Obtain purchasing information which shows purchasing habits over the past two years for the conveyor rollers. Look for a change in materials and/or vendors.	Robert Tomlin	5/6/2001	No	
Parts	Failed Bearings	Collect past, present and future failed bearings. Take to machine shop and cut open for inspection.	Robert Tomlin	5/6/2001	No	
Parts	Lubricant	Collect lubricant samples from failed bearings and from lubricant storage facility. Take samples to lab for testing and comparison.	David Craig	5/6/2001	No	
People	Lubricators	Interview lubricators about how often they lubricate, what type of lubricant they use and how much lubricant they apply.	Bob Latino	5/6/2001	No	
Position	Position of Failed Rollers	Map out locations of failed bearing and see if they are isolated to one area or sporadic in location.	Bob Latino	5/6/2001	No	

FIGURE 13.14 Sample Data Strategy Report.

Base and Additional Costs

Description	Cost
Manhours(3)	$30.00

Total Cost : $30.00

Save & Close

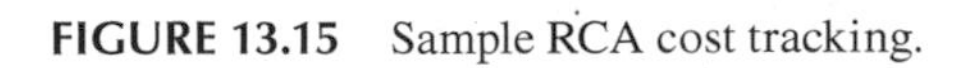

FIGURE 13.15 Sample RCA cost tracking.

AUTOMATING THE ANALYSIS TEAM STRUCTURE

We discussed at length the importance of the team structure. If you remember, we discussed how important the diversity of backgrounds is to a successful result. We also stressed that the leader of an RCA team should typically not be the expert in the event being analyzed because of the inherent bias that may persist.

We discussed the focus of the team structure by formalizing the team entity through the development of a Team Charter and the identification of Critical Success Factors (CSFs). These tasks show management that considerable thought was invested about why the team was formed and what its objectives are in obtaining success.

Now let's contrast the manual version with the automated version. In a manual format we would most likely be utilizing a paper filing system to record team member information. We would also likely use a word processing program to develop the Team Charter and the CSFs. With an automated format, we can use PROACT to catalog all this information in one location along with the 5P's information collected previously.

If you will remember, all of this was collected using the New Analysis Wizard (nine steps) when we created the new analysis. The Order the Analysis Team tab is merely where team information is stored and available for modifications. In the example in Figure 13.16 we show a change in granting permissions to a team member. The Principal Analyst easily completes this task by simply double-clicking on the team member's name.

PROACT will maintain a Team Pool in which a database of qualified RCA team members is stored. "Qualified team" members may be former RCA participants, individuals who have received RCA training in the past, or individuals who posses a certain expertise that is difficult to find. In any case, maintaining a record of such talent is an efficient way of helping organize RCA teams. Once a reservoir of talent has been identified, specific individuals can be assigned to lead and participate on the core team. These choices will obviously vary based on the nature of the event

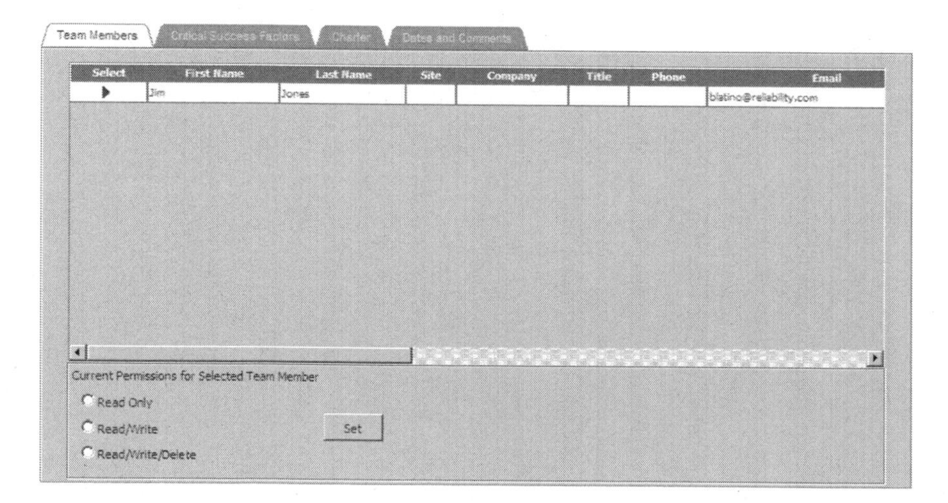

FIGURE 13.16 Setting team member permissions.

being analyzed. PROACT will allow reports to be developed on the team members based on their names and/or telephone numbers.

PROACT now has all the team information cataloged and organized within a database. Up until this point, there has been no need to utilize individual database or spreadsheet programs, or word processing programs. It is all located within one RCA file.

AUTOMATING THE ROOT CAUSE ANALYSIS— LOGIC TREE DEVELOPMENT

Moving on, let's assume we are at a point in the RCA where the initial data needs have been identified, assigned, and collected, and an ideal team has been put together and organized to approach the task of analysis. Now we face the real issue of analyzing the data to determine what happened.

Using the manual method (paper based) to develop the logic tree has its pros and cons. One of the disadvantages is double handling of data. In the manual method, a logic tree is built in a conference room where a *mural* has been put together made of easel pad paper or craft paper. Subsequently the analysts will facilitate the team using Post-Its®.* This means that at some point in time, this information will have to be transcribed into another format for inclusion into the report and/or displayed in a presentation. This double handling leads to an inefficiency of time as well. When a team meeting ends, the team members usually do not have the updated logic tree until days later. This results in unnecessary delays before all team members have consistent information.

One of the psychological advantages of using the manual method in conjunction with the automated method is it can be perceived as accomplishing work. We have seen the paradigms at play where many believe if someone is working on a computer all the time that work is not being accomplished. Some may feel that if wrenches are not being turned or machines are not being operated, then work is not being accomplished. The same can be said for RCA. If management walks by a conference room where an RCA team is meeting and only sees one laptop on the table and five team members sitting around talking, then it can be perceived as a non-value-added use of time. However, in the same scenario, if management walks by and sees this huge piece of craft paper on the wall with all these Post-Its®, it can be deemed as tangible work (even if a recorder has duplicated the logic tree within PROACT on the laptop within the same meeting).

From an efficiency standpoint, using a laptop and an LCD projector in a team meeting is the ideal forum to conduct logic-tree-building sessions. This will obviously have to be the determination of the analyst or the team based on the resources they have available to them at their site.

We will now go through how PROACT can help automate the analyzing of data. PROACT was developed using the same logic rules as discussed in the RCA Method described in this text.

---

* Post-It is a registered trademark of the 3M Corporation.

The opening screen in ANALYZE is basically a blank worksheet with the necessary tools or tabs to build the logic tree. When building the logic tree during a team meeting, the analyst should start with the Top Box Wizard tab. This will prompt the team to enter the exact event they are ultimately analyzing. The detailed descriptions of how to develop events and modes are located in Chapters 5 and 10. Once an event has been entered, the team will then be prompted to enter the various modes that apply under the circumstances. Only enter as many modes as are necessary for the particular event being analyzed (Figures 13.17 to 13.19).

Visualize as we go through these scenarios that the LCD projection is on a screen and the entire team can view the building of the logic tree as it is being developed. At the same time, it is being recorded in the PROACT RCA file in the software. If

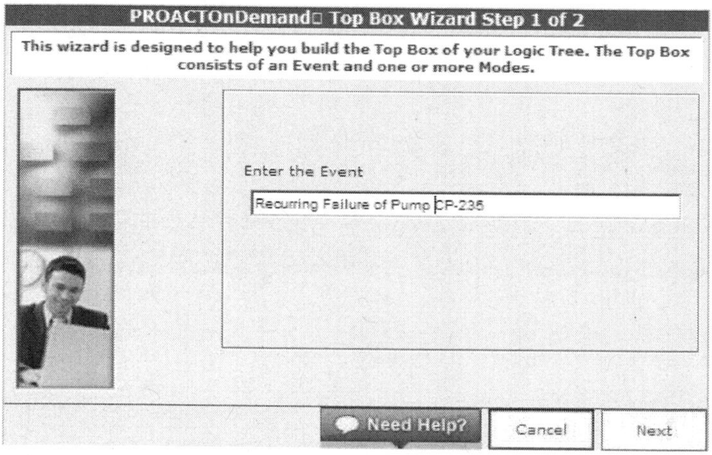

FIGURE 13.17 Top Box Wizard/Event, Step 1.

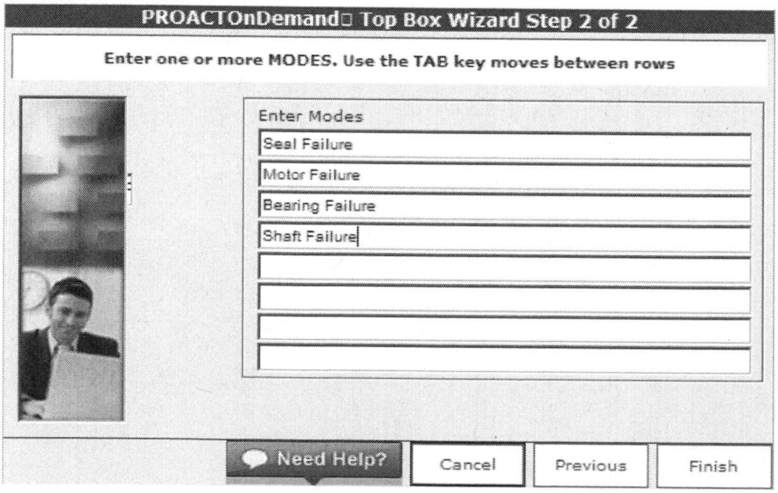

FIGURE 13.18 Top Box Wizard/Modes, Step 2.

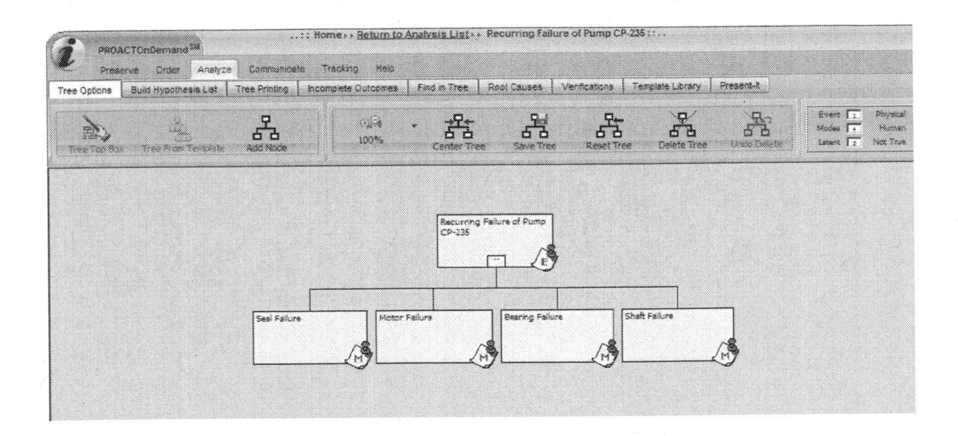

FIGURE 13.19 The completed Top Box.

the Top Box has been outlined, then the known facts of the situation have been identified and we must begin the process of hypothesizing *how could* these facts have occurred. This is where the analyst plays the role of a facilitator and begins the continual questioning process of "How Could?" the preceding event have occurred. The core team of experts will be the source for the answers from various perspectives. As appropriate hypotheses are thrown on the table, they are entered into the logic tree. As each hypothesis is entered, the user will be reminded of the need to fill out the verification log. The verification log form will pop up every time a hypothesis is entered and will

1. Prompt the team to designate who will be responsible for testing the hypotheses,
2. List what test is to be performed,
3. List the anticipated completion date of the test,
4. List the actual completion date of the test, and
5. List the test outcomes and the confidence level that the hypothesis is true or false based on the test.

Remember in Chapter 10 ("Analyzing the Data") we discussed the maintaining of a verification log manually. Again, the PROACT software is just another glorified electronic data collector for organizing the verification data (Figure 13.20).

Once a verification task has been assigned a person responsible, a verification method, or a test to be done and a completion date, it is logged in and stored to await an outcome. Sometimes such verifications will require the attachment (file linking) of the proof of the verification test. This can be in the form of test results, pictures, reports, procedures, etc., and can be accomplished in the same manner described previously in the section on preserving event data. Once again, this is the second location in PROACT where we can assign a task, so we can also log in the time it took to do the task and issue e-mail reminders.

As the reiterative process moves forward and more and more hypotheses are developed, the logic tree continues to grow. The Tree Objects provides the user

Verification Log

Verification Log

Verification Method

Exp. Completion Date

07/15/2010

Verification Outcome

Completion Date

08/14/2010

File Links and Options

Add... View... Remove...

Hours Expended

Team Member

Bob Latino

View/Enter Additional Costs

Save Changes to the Verification

Cancel Changes

In the Verification, you may add files to each Data Collection Record to support your RCA. As you add files, they will be listed above. Almost all file types are permitted as long as they do not exceed 20MB Each.

Add: This button allows you to select from a list of previously uploaded files for this RCA, or upload new files. After uploading you may select the new file for insertion to above.

View: To view an existing file above click the blue arrow. You may view and or download it.

Remove: Select the file you wish to remove. Click remove. This will remove the selected file from this record.

Toll Free: 1 800 457-0647

FIGURE 13.20 Sample Hypothesis Verification Log screen.

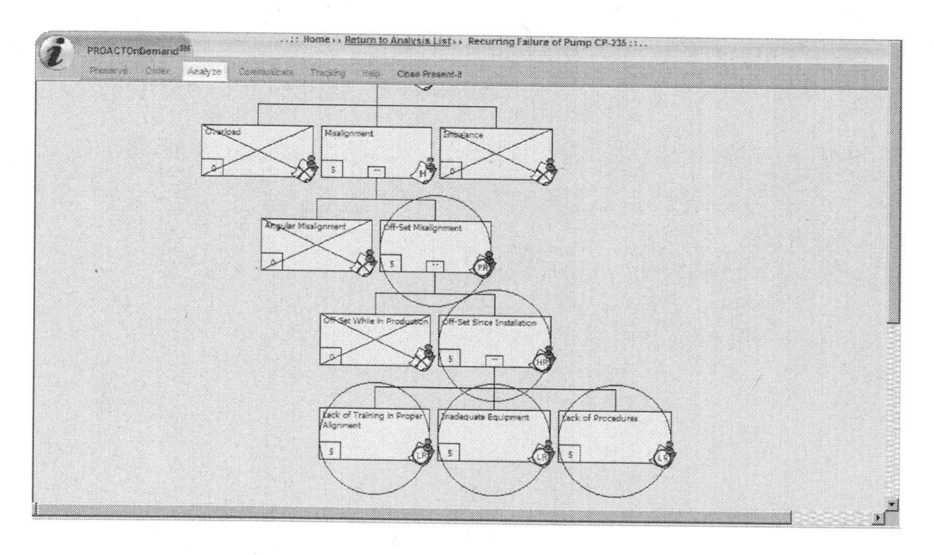

FIGURE 13.21 Sample Root Causes.

various options for adding hypotheses to the logic tree as well as manipulating the tree (i.e., zoom, center, etc.). When the team identifies Physical Roots (PR), the analyst can simply place the cursor on the hypotheses the analyst wishes to identify as a PR and right click on the node, and the menu options will appear (Figure 13.21). Select RELABEL AS and choose the appropriate root type. Consequently, in the background of the program, this root cause will now automatically come up in the report writing section and require a recommendation to be made as a countermeasure.

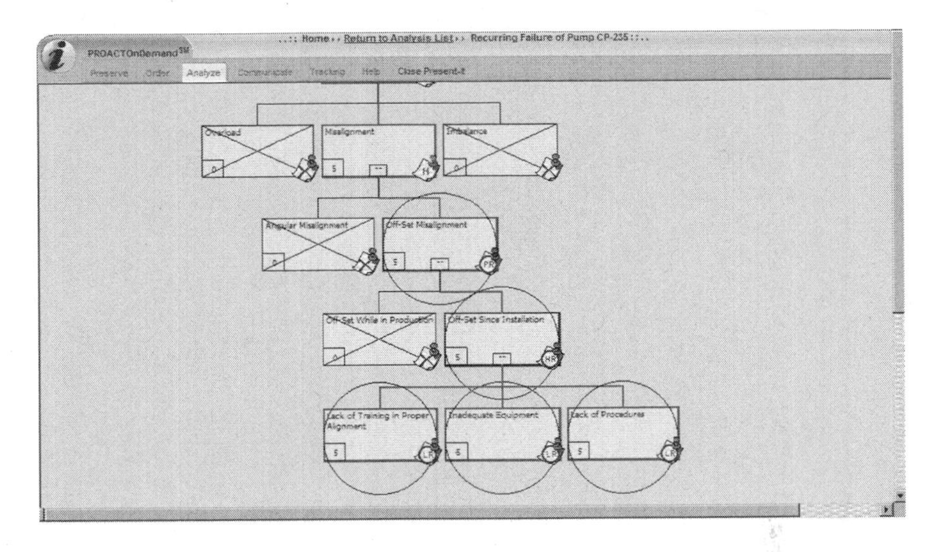

FIGURE 13.22 PROACT's Path to Failure option.

At this point in the RCA, considerations will begin in regard to the final presentation of the logic tree to management. To this end, PROACT provides a presentation mode to eliminate the need for developing a presentation using a graphics program in another separate file. By clicking the "Present-It" icon on the node menu, a full-screen presentation mode will appear with the entire logic tree expanded. This mode will allow the team to make its final presentation in real time. The speaker can begin with the collapsed logic tree showing only the event and modes. Then, as the presentation progresses, the speaker can expand on the hypothesis in question and show the possibilities. If a manager questions any hypothesis, all the speaker has to do is right-click on that block and select Verification Log. The Verification Log pops up to show how the hypothesis was tested and what the result was. This is an extremely useful feature when making a presentation. In the end, the analyst can activate the "path to failure" feature and only the logic tree paths with identified root causes will be highlighted (Figure 13.22).

PROACT's logic tree provides a unique feature to allow analysts to capitalize on the successful logic of past analyses. The feature entitled "Previous Suggestions" allows the team to search all past published analyses for instances in which certain similar words were used in the tree. The unique search feature will also scan any published templates input by either the software provider or the facility itself.

For example, if we knew we had a bearing failure in our situation, and we knew the question was "How could a bearing fail?" we could activate this feature to see how others have answered this in the past. In this case, PROACT would search all past, published analyses and any applied templates that had the word "bearing" in their trees, and things like erosion, corrosion, fatigue, and overload may appear. The analyst can then determine if any are applicable and select which ones to add to the logic tree (Figure 13.23).

Some automated systems on the market provide pick lists, which lead users to believe that all the available options to answer their questions are embedded in the list.

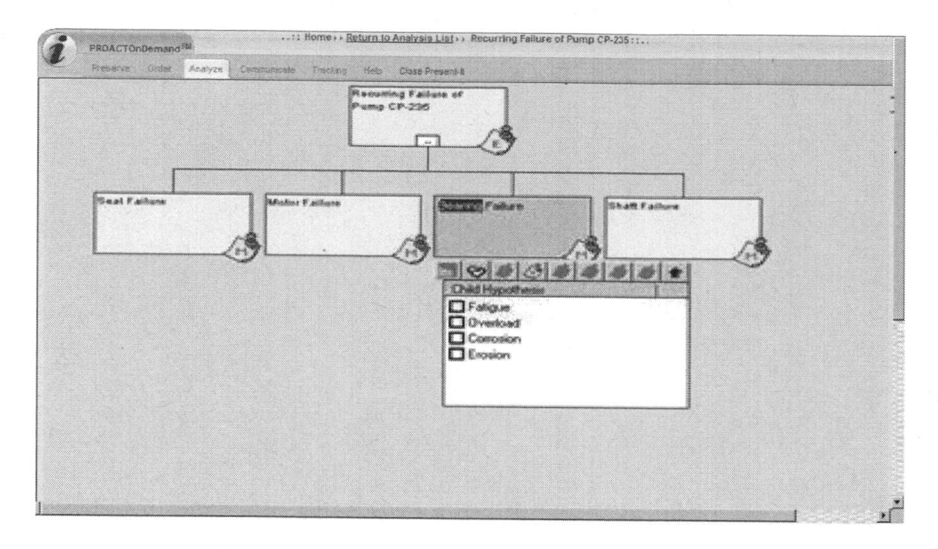

FIGURE 13.23 Sample of Previous Suggestions feature.

This will never be, as all the possibilities cannot be captured that cover all of the potential variables at play. Humans, being prone to the path of least resistance, often will abuse such systems and pick the closest answer they feel is the case. This is dangerous and is often referred to as "RCA by the numbers." We purposely named this feature "Previous Suggestions" to reflect the experience of specific facilities. We encourage facilities to input past analyses into PROACT to get the knowledge base started.

RCI also put in its own experience-based templates to reflect the experience of over hundreds of mechanical, electrical, and administrative type events. These can be loaded into PROACT and used under the Previous Suggestions feature immediately.

The PROACT software was purposely developed to be compatible with almost all competing methodologies on the market. Analyses from those methods can be easily input into the PROACT software and manipulated in various fashions to produce more information and to express in lessons learned.

Considering the time it takes to develop a formal presentation and the subsequent supporting data such as verification information, PROACT considerably reduces the time to perform such tasks. Again, thus far, all information that is related to this RCA is still located in one file using one program.

AUTOMATING RCA REPORT WRITING

One of the most tedious tasks in conducting a full-blown RCA is the writing of the report. If no standard formats are available, then this can be a laborious task that lacks continuity. Without standard formats, the consistency of reporting results suffers and the information is ignored or not understood. In the "manual" method of writing reports, we would generally use a word processing program and develop a stand-alone report with a table of contents that suits the team. Then someone, usually

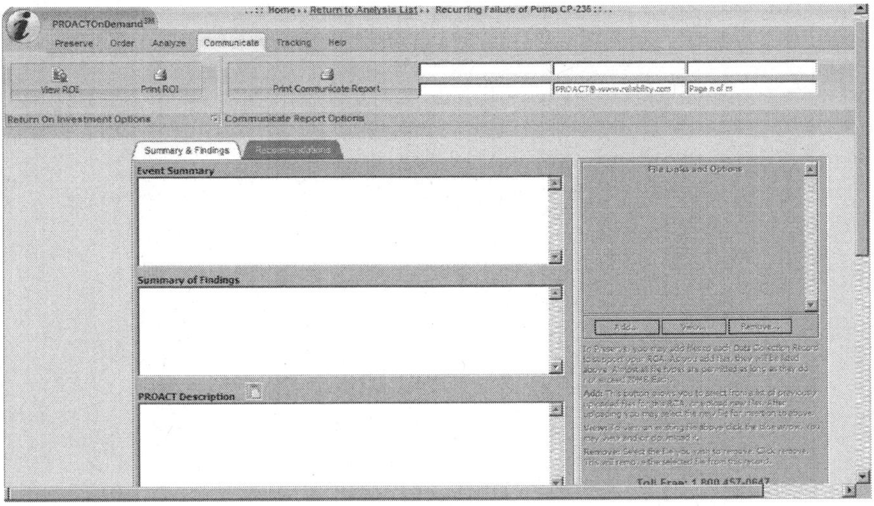

FIGURE 13.24 Communicate—Report Summaries screen.

the Principal Analyst (PA), is charged with the task of developing the content and typing it into an acceptable format. While the team members may contribute, the brunt of the legwork is on the shoulders of the PA. Then the report must be properly distributed to the parties that would benefit the most. All in all, the task is extremely burdensome and is not the highlight of the analysis work.

PROACT provides analysts with a report writer where the authors can report only the topics they wish to those they wish to see it. The customized report feature breaks the report into three sections:

1. The Summaries—Event and Findings Summaries
2. The Recommendations—Executive and Detailed Recommendations
3. The Custom Table of Contents—Supporting Data in a Desired Format

Each of these sections was discussed at length in Chapter 11. Our purpose here is to show how the report-writing task can be automated (Figure 13.24). Within the Summaries fields we are prompted to fill in an event summary and findings summary and the PROACT process description (or other process that may have been used).

As we also discussed in Chapter 11, the entire RCA process revolves around the final presentation and getting recommendations approved. The Root Cause Action Matrix was the culmination of the entire analysis. To this end, PROACT provides such a matrix, which requires input in various fields. Any hypothesis on the logic tree that was designated as a root cause in the ANALYZE section will automatically appear in the drop box along with the appropriate type of root cause it was identified as (physical, human, or latent). PROACT will also seek a person on the team (or someone else) as being responsible for implementing the recommendation by a certain date. Therefore, when the logic tree has been completed, the roots,

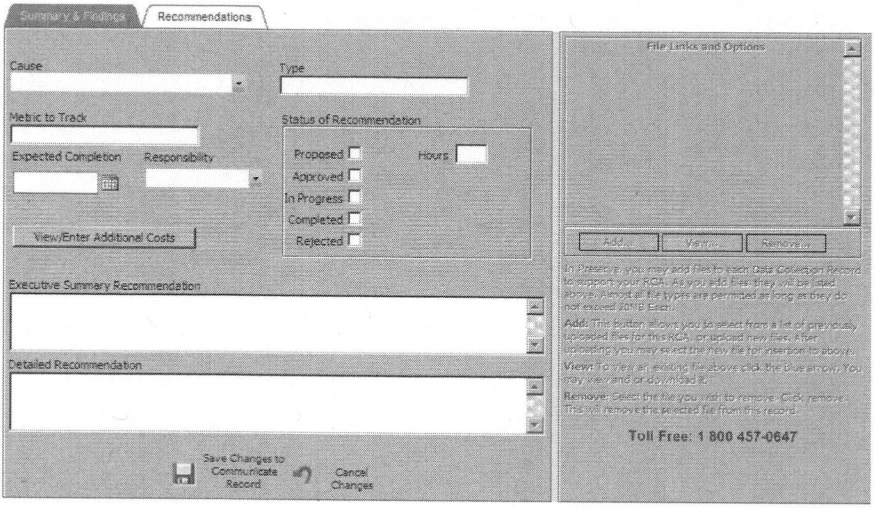

FIGURE 13.25 Analysis recommendations.

which require recommendations, should be assigned to various individuals and they should set target dates by which to complete them. In this section, we put in all the information related to both the executive summary and detailed recommendations. This information includes responsibilities for development and implementation, dates, disposition of proposed recommendation, metric to track, and cost information (Figure 13.25).

Of course, all of this effort is for naught if you cannot print the report. PROACT allows the author to print the entire report, or to select sections of preference only. A Print Wizard allows the author to customize covers and also to print selected topics, if desired. This automating of the report writing means that formal reports do not have to be developed from scratch; we do not have to worry about formatting or standardizing because PROACT does it all automatically.

During the development of the OnDemand Version 3.0 of PROACT, almost all of the feedback from our Enterprise Version 3.0 users was incorporated. One of those changes was in the reporting section. Many users understood that PROACT was an acronym where the steps in the process were sequential. However they preferred the option to change the sequence of the steps in the process if they wanted so they could set their own table of contents in the report. As a result we added a Print Wizard feature that allowed them to do just that—customize their table of contents.

By selecting the information we desire to be in the report and in the order we select, we can customize our table of contents (Figure 13.26). This is significant because we may report externally (i.e., to regulators) as well as internally (i.e., to the CFO). PROACT will allow the analyst to store different Table of Contents (TOCs) for such purposes.

The final step in the Print Wizard is the print preview (Figure 13.27). This allows the user to preview and scroll through the entire report. Understanding that we are operating in a true web environment, the output report will be in .pdf format.

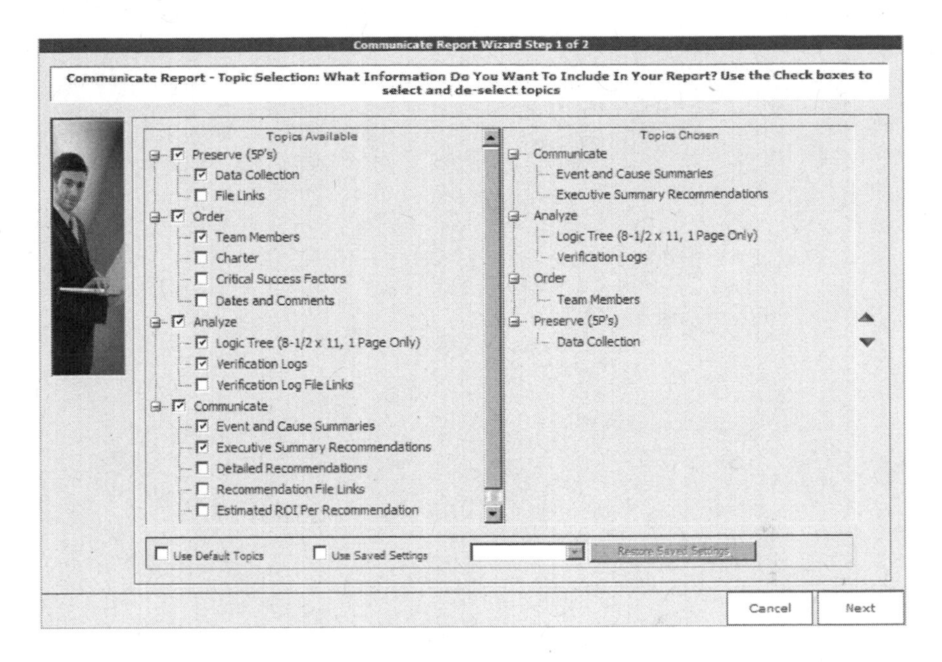

FIGURE 13.26 The PROACT Print Wizard.

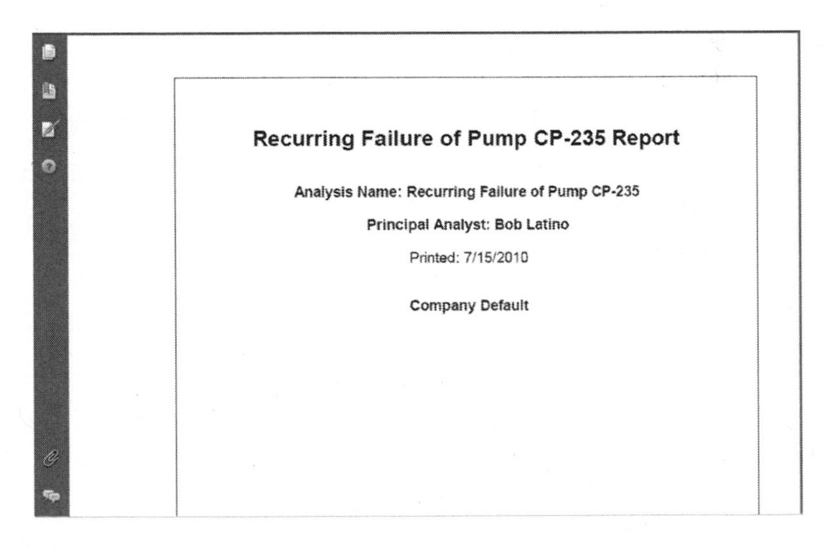

FIGURE 13.27 The PROACT print preview.

AUTOMATING TRACKING METRICS

As we know, we are not successful at RCA unless some bottom-line metric improves. Therefore, we must select and monitor over time the metric of choice. In a manual format we may have to be diligent about getting certain data from certain reports, or we may have to develop a whole new report to get the information we seek.

One thing we should not do is make the tracking process so complicated that it is too difficult and frustrating to accomplish. PROACT was designed to make this tracking process very simple, basic, and user friendly (Figure 13.28). Tracking also has its own 4-Step Wizard, which will walk the user through a series of questions, such as

1. Save Graph as: ____________
2. Title of Graph: ____________
3. Sub-Title: ______________
4. Tracking Intervals: ________
5. Tracking Periods: _________
6. Tracking Metric: __________
7. Data to Input: ____________

This provides enough data to make an easy-to-follow, basic graph. Each month when new data is available it can be input into the Wizard to easily update the graph. If analysts were using PROACT in an enterprise environment that is integrated with other data systems such as CMMS, this data could be collected automatically. Oftentimes we are questioned about whether or not PROACT can import data from other systems, as the data sought is already available elsewhere. For instance,

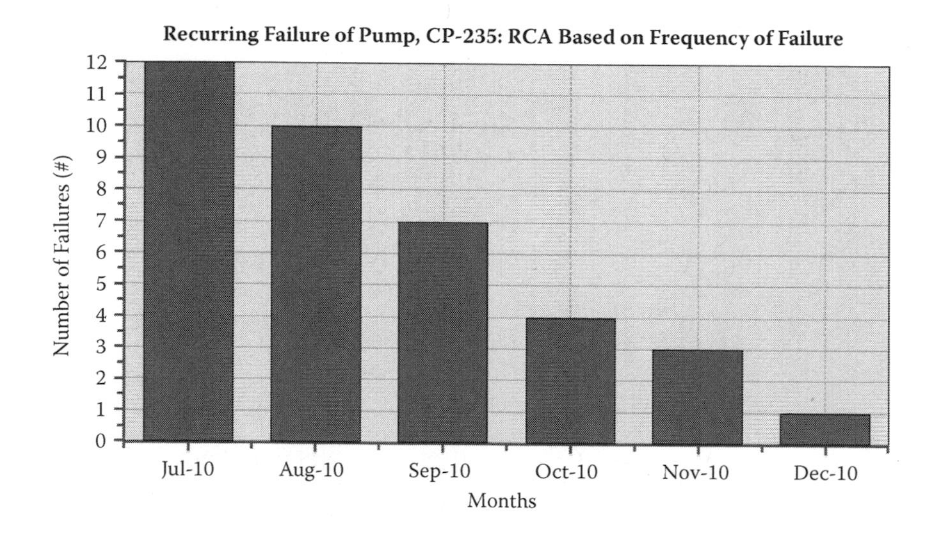

FIGURE 13.28 The PROACT Bottom-Line Tracker.

Select	Analysis Name	Analysis Description	Analysis Type	Owner	AnalysisID	Analysis Cost	Start Date	Expected Completion Date
	Endobronchial Fire		Risk	Bob Latino	1	$0.00	12/18/2009	01/18/2010
	Allergic Reaction		Claims	Bob Latino	2	$0.00	01/29/2010	02/26/2010
	Failure of CP-235		Operational	Bob Latino	3	$0.00	01/15/2010	02/15/2010
	Customer Complain	Left-Click to Select Analysis/Right-Click for additional Analysis Options	Mechanical	Bob Latino	4	$0.00	01/19/2010	02/19/2010
	Conveyor Roller Failures		Operational	Bob Latino	5	$0.00	01/19/2010	02/19/2010
	Converter Failures		Mechanical	Bob Latino	6	$0.00	01/19/2010	02/19/2010
	Medication Reconciliations Errors		Mechanical	Bob Latino	7	$0.00	01/26/2010	02/26/2010
	Injury from Falls		Environmental	Bob Latino	8	$0.00	01/28/2010	02/28/2010
	Cuts and Lacerations Injuries		Environmental	Bob Latino	9	$0.00	02/01/2010	03/01/2010
	Environmental Release		Administrative	Bob Latino	10	$0.00	04/09/2010	05/09/2010
	OSHA Recordable		Mechanical	Bob Latino	11	$2,500.00	04/21/2010	05/21/2010
	EPA Violation		Operational	Bob Latino	12	$0.00	05/11/2010	06/11/2010

FIGURE 13.29 PROACT Publisher.

PROACT can import Team Member data from a program like MS Outlook.* However, other proprietary programs like Computerized Maintenance Management Systems (CMMS) may not be so cooperative in opening up their tables to other vendors. A database is a database so all you need is the cooperation of the database owners.

The development of a dynamic tracking graph completes the circle of finalizing an RCA. Automating this graphing feature in PROACT alleviates the need to use a separate graphics package to make the graph.

Think back now—if we use the traditional manual method, we would require the use of a database package, a spreadsheet package, a word processing package, and a graphics package in order to complete the RCA. This would require the alignment of file names and so on for continuity. PROACT compiles everything in one location and the file can be e-mailed to others to ensure proper distribution.

PROACTOnDemand℠ allows for the efficient and effective knowledge transfer of successful analyses to others in the company who may benefit. PROACT puts information at the fingertips of those who can use it most. Until an analysis is completed, only the Principal Analyst and the team members can see their work in progress. However, when they are done and the Principal Analyst certifies that the analysis as complete, everyone who has permission and access to PROACT will be able to view the results in a "read only" format. This feature is called Publishing (Figure 13.29). Once an analysis has been published, its icon within the database changes (a notebook icon appears) allowing us to visually recognize those analyses that are completed. Also, ONLY completed analyses can be searched based on various criteria.

These features, as well as many others, allow analysts to focus more of their time on doing the analysis rather than on the administrative tasks to document the process and transfer the knowledge. PROACT is truly a proactive tool when conducting RCAs.

* MS Outlook is a registered trademark of the Microsoft Corporation.

It has been our experience that utilizing PROACT to facilitate RCAs in the field has reduced the administrative time to complete them by approximately 50%. This means that, from a productivity standpoint, analysts can complete more analyses in a given time period if they automate their RCA processes.

PROACT was presented a Gold Medal Award (general maintenance software) in *Plant Engineering's* "Product of the Year." For more information about how to obtain PROACT contact

Reliability Center Incorporated
501 Westover Avenue, Suite 100
P.O. Box 1421
Hopewell, VA 23860
Telephone: 804-458-0645
Fax: 804-452-2119
Web address: http://www.reliability.com

14 Case Histories

This chapter puts into practice what this text has described in theory thus far. We have described in detail the Root Cause Analysis (RCA) method and provided some academic examples to further your understanding of the concepts.

The following case studies are a result of having the right combination of management support, the ideal RCA team, and proper application of the RCA methodology. RCI commends the submitters of these case histories for their courage in allowing others to learn from their experiences. These corporations and their RCA efforts have proven what a well-focused organization can accomplish with the creative and innovative minds of its workforce.

As you read through the summaries of these actual case histories, you will notice that the Returns on Investment (ROIs) for eliminating these chronic events are expressed in the thousands of percent. Had we not had permission to publish these remarkable returns, would anyone have believed they were real? You will also notice that the time frames to complete the RCAs ranged from days to months. While these results are without a doubt impressive, they are easily attainable when the organizational environment supports the RCA activities. Read on and become a believer.

CASE STUDY NO. 1: NORTH AMERICAN PAPER MILL

Undesirable Event: Repetitive Thick Stock Pump Failures

Undesirable Event Summary: During the years of 2007 and 2008, there were 19 failures on thick stock pumps (Figure 14.1) on A and B unit in the Bleach Plant. Several attempts had been made to implement corrective actions for the pumps, but ultimately failures were still occurring. Thick stock pumps are big-ticket items ranging from $60,000 to $120,000 per rebuild due to the tight clearances and amount of material it takes to machine the pumps. It was determined by maintenance that the pumps could be rebuilt in-house in the bleach room maintenance shop. This has been very successful and has cut the cost of maintenance dramatically and has proven to provide greater reliability. Performing the rebuild by in-house millwrights has brought ownership and pride to the repairs and operations of the thick stock pumps. Although production loss was not used in the Opportunity Analysis for these failures, it would have been a significant factor in the loss equation for these events.

Through the RCA, the team determined that the two prevalent causes for premature failure were bearing failures and foreign material being introduced into the pumps. Foreign material is an issue that still persists and is a random failure. This occurs because the clearance (0.03) in the pumps is very small and does not allow any metals bigger than a paper clip to be passed through (Figure 14.2). With closer attention to this issue, there have not been any failures due to foreign objects in the stock reported since this analysis was completed.

FIGURE 14.1 Thick stock pumps.

FIGURE 14.2 Cross-section drawing of thick stock pump.

TABLE 14.1
Line Item Opportunity Analysis

Subsystem	Event	Mode	Frequency	Impact	Total Annual Loss
Bleach Room	Failure of Thick Stock Pumps	Bearing Failures	10 per year	~$100,000	~$1,000,000

The RCA team focused on bearing failures during the RCA, which led to many preventive actions that were recommended and implemented. Several issues had been identified that caused contamination to find its way into the lubrication chambers. Packing was the first area that was pursued. New flow visual indicators and check valves on the packing water lines helped to ensure sufficient packing water was getting to the packing. A new procedure was recommended to properly pack a thick stock pump and was later implemented. To monitor how often the lubrication becomes contaminated, a predictive maintenance technique (lube oil analysis) is now being utilized. Oil analyses are being performed on all thick stock pumps every 2 weeks in order to determine how often the oil requires changing. This will be sustained by putting it on the area oiler's checklist.

Also, the team implemented both operator and maintenance ECCM routes (essential care/condition monitoring) to monitor the operation of these critical pumps. Operators perform visual inspections of the thick stock pumps and motors, in addition to recording vibration, temperature, and pressure readings into handheld data collectors for early detection of defects. This is in addition to other more advanced vibration analyses that take place (Table 14.1).

Identified Physical Roots:

- Lubrication issues due to contamination
- Pack failures
- Packing water failures (plugged, inadequate pressure, etc.)

Identified Human Roots:

- Inadequate packing
- Inspections not performed
- Poor rebuilds from outside services

Identified Latent Roots:

- Using the wrong packing procedures
- No formal inspection routine for operations and maintenance
- Inability to determine if packing lines were in working order
- Assumed that repairs were performed properly by OEM

Implemented Corrective Actions:

- Improved packing procedure implemented and documented in SAP PM.
- Oil Analysis implemented.

- Development and execution of ECCM inspection routes.
- New visual flow indicators installed for packing water lines (Figure 14.3).
- Improved check valves installed for all thick stock pumps (Figure 14.4).
- Defined all new parts as storeroom items for future use.
- Perform in-house rebuild on thick stock pumps.

FIGURE 14.3 Thick stock pump Visual Flow Indicators.

FIGURE 14.4 Thick stock pump improved Check Valves.

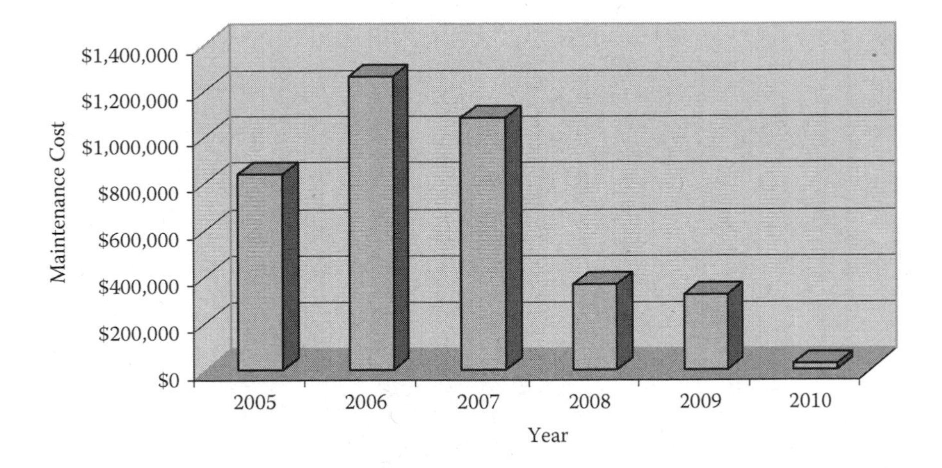

FIGURE 14.5 Thick stock pump maintenance costs.

Effect on Bottom Line:

Tracking Metrics

- Mean-Time-Between-Failure (MTBF) increased from approximately 6 months to over 2 years.
- Maintenance cost reduced less than 25% of pre-RCA costs.

Bottom-Line Results

- There are substantially fewer failures on these critical pumps, and defects are caught before they cause catastrophic problems.
- In-house rebuild of pumps has resulted in greater ownership and pride in the performance of the pumps.
- Maintenance costs in 2010 YTD (July) are approximately $25,000 compared to roughly $500,000 in 2006 and 2007 in the same time frame (Figure 14.5).

RCA Team Statistics:

- Start Date: 7/30/2008
- End Date: 10/20/2008
- Estimated Cost of Performing the Analysis: $20,000
- Approximate Savings: $700,000 per year
- Estimated ROI: 3,500%

Corrective Action Time Frame:

- Most of the corrective actions were implemented in less than 6 months.
- The analysis spanned about 3 months.
- The implementation of corrective actions was complete approximately after the completion of the analysis.

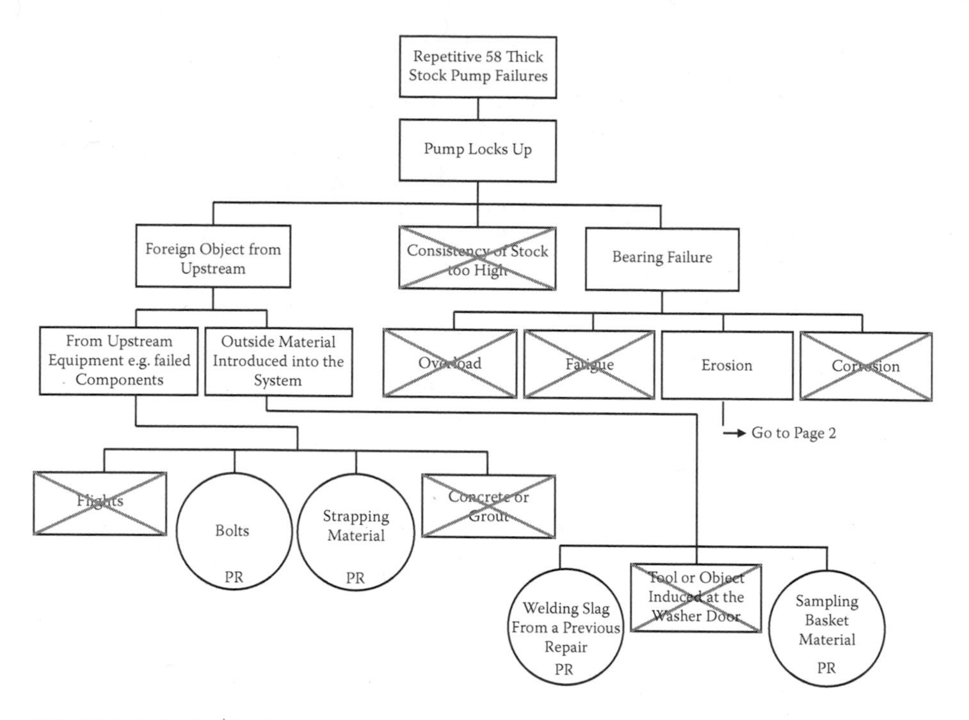

FIGURE 14.6 Logic Tree.

Core Team Members:

- Arnie Persinger (Principal Analyst)
- Dean Muterspaugh
- Todd Fix
- Robert Newcomer
- Josh Taube
- Rob George
- David Persinger
- Will Sales

Special thanks to Arnie Persinger (Area Maintenance Superintendent) and Craig Lane (Fiberline Operations Superintendent) for believing in the process and ensuring a successful outcome.

CASE STUDY NO. 2: PEMMAX CONSULTANTS, WATERLOO, ONTARIO, CANADA

Undesirable Event: Low MTBF of Cooler in Leaching Operation
Undesirable Event Summary: In a leaching operation, titanium shell-and-tube heat exchangers are used to cool from 250°C to 130°C a water-based slurry with 30% mineral solids and a pH = 1; the plant uses 10 of these coolers and their average life was 20 days for annual production and maintenance losses of approximately $11,500,000.

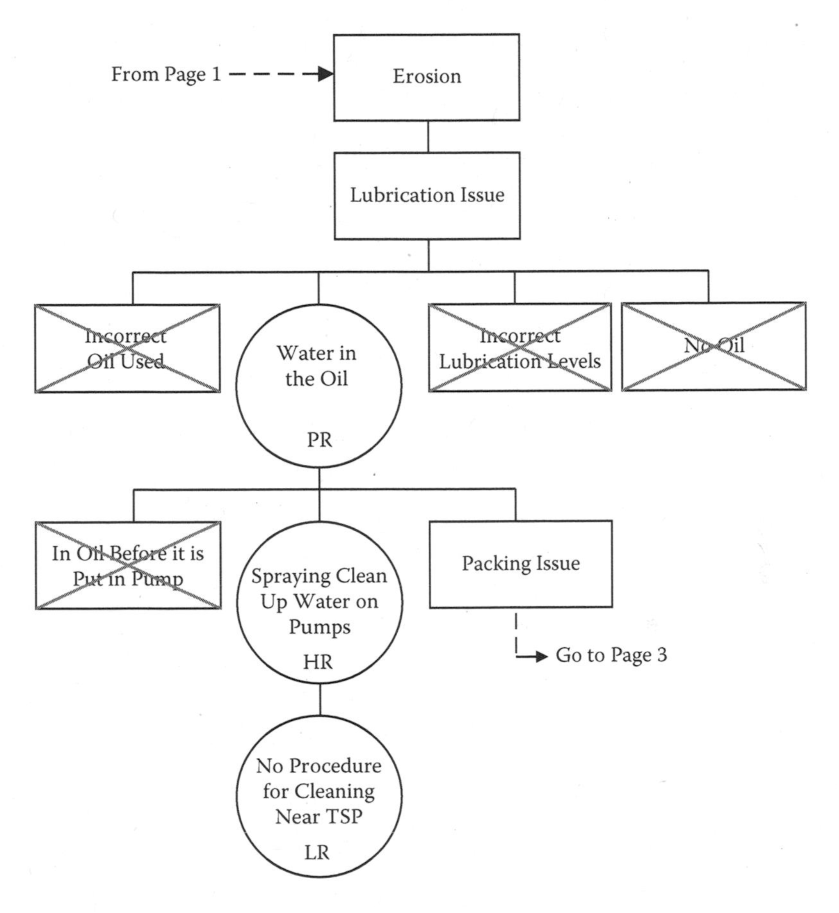

FIGURE 14.6 (continued).

When leaching became the production bottleneck, a cross-functional team (six employees led by Tony Rodriguez of Pemmax Consultants) was given the challenge of increasing the average operating life of these units from 20 to 40 days.

Line Item from Opportunity Analysis Related Total Annual Cost of Described Event

Identified Physical Roots:

- Large crust pieces in the flow stream obstructed tube inlets.
- Dead area in feed head to coolers.
- Abrupt (not tapered) inlet to cooler tubes.

Identified Human Roots:

- Maintenance engineers did not investigate the problem deeply enough.

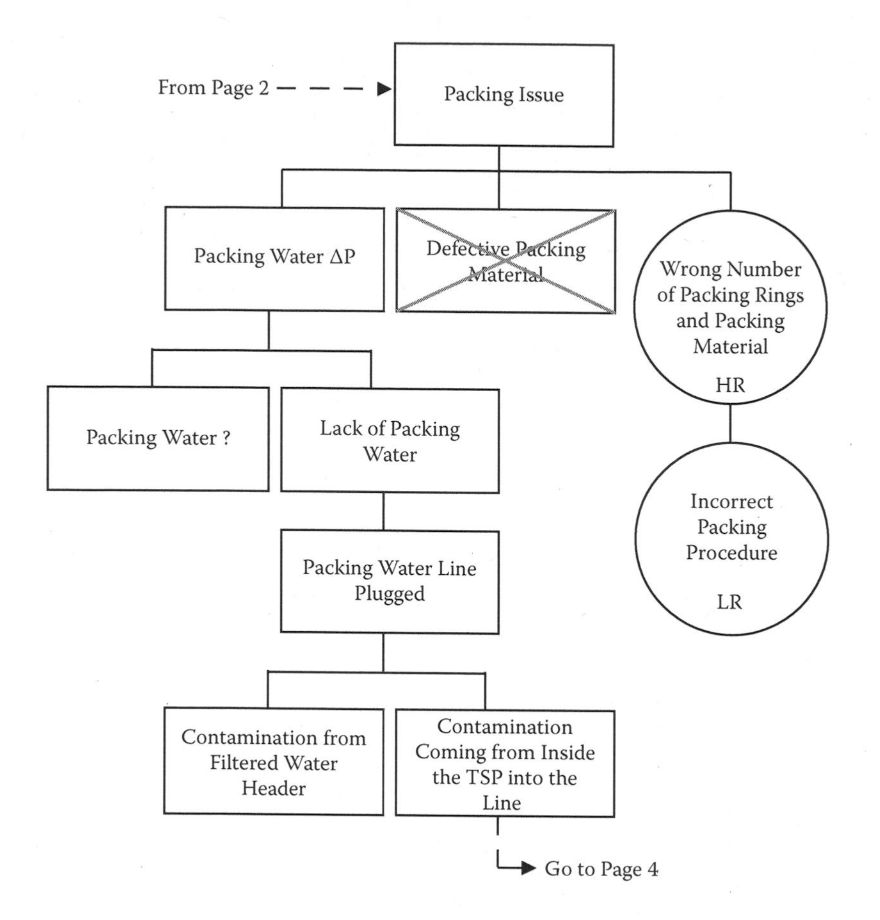

FIGURE 14.6 (continued).

TABLE 14.2
Line Item Opportunity Analysis—PEMMAX

Event	Mode	Frequency/ Yr	Manpower $/ Occurrence (MP$)	Materials/ Occurrence (MATRL$)	Lost Profit Opportunity (LPO$)	Total Annual Loss
Cooler failure	Weld erosion	130	$6,000	$12,000	$70,000	$11,440,000[a]

[a] Indicates this is where the derivation of the total loss came from.

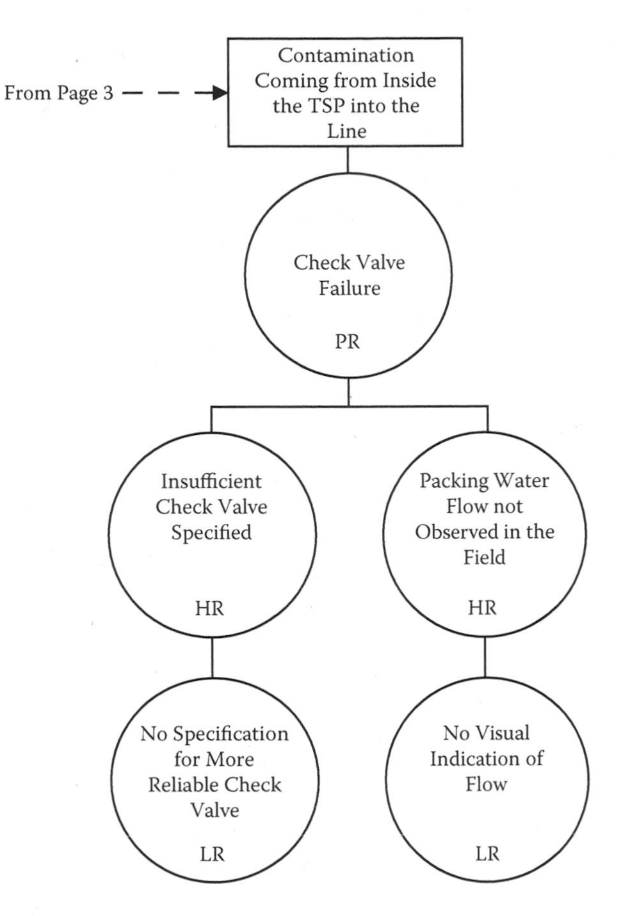

FIGURE 14.6 (continued).

Identified Latent Roots:

- Previous administrations did not encourage comprehensive analysis of problems (RCA was foreign to them).

Implemented Corrective Actions:

- Filter out large crust pieces.
- Eliminate dead zone in the feed head (Figure 14.7).
- Weld tapered inlets to cooler tubes.

Effect on Bottom Line:

- MTBF increased from 20 to over 60 days.
- Estimated savings as a result of the RCA were calculated at $7.6 MM/yr or just over $600,000/month. Savings are two-thirds of the initial losses; remember, we tripled the MTBF of the coolers.

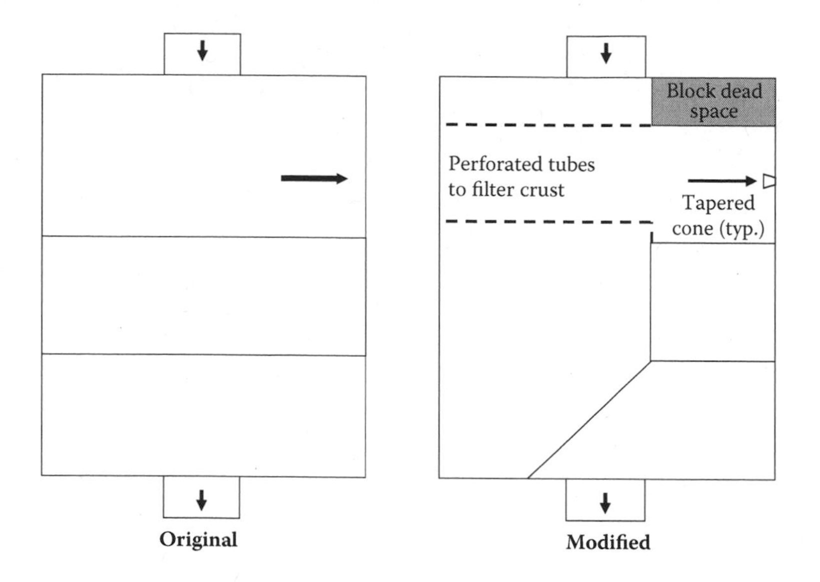

FIGURE 14.7 Cooler Feed Head modifications.

Additional Comments:

- Consulting plus labor costs of company employees: approx. $300,000
- Payback at current savings less than 2 weeks

RCA Team Statistics:

- Start Date: November/2005
- End Date: December/2006
- Estimated Cost of Performing the Analysis: $300,000
- Approximate Savings: $7.6 MM/yr
- Estimated ROI: >2,500%

Corrective Action Time Frame:

- The time frame from the time that the event occurred to the time the corrective actions were taken was 1 year.

CASE STUDY NO. 3: PSEG, JERSEY CITY, NEW JERSEY

Undesirable Event: Pulverizer Explosion

Undesirable Event Summary: In a power plant, Pulverizer operations are designed to feed the boiler by grinding and conveying pulverized coal with air at desired ratios in order to maintain efficient boiler combustion during steam production (Figure 14.9). Due to the highly combustible atmosphere of the system it is possible to

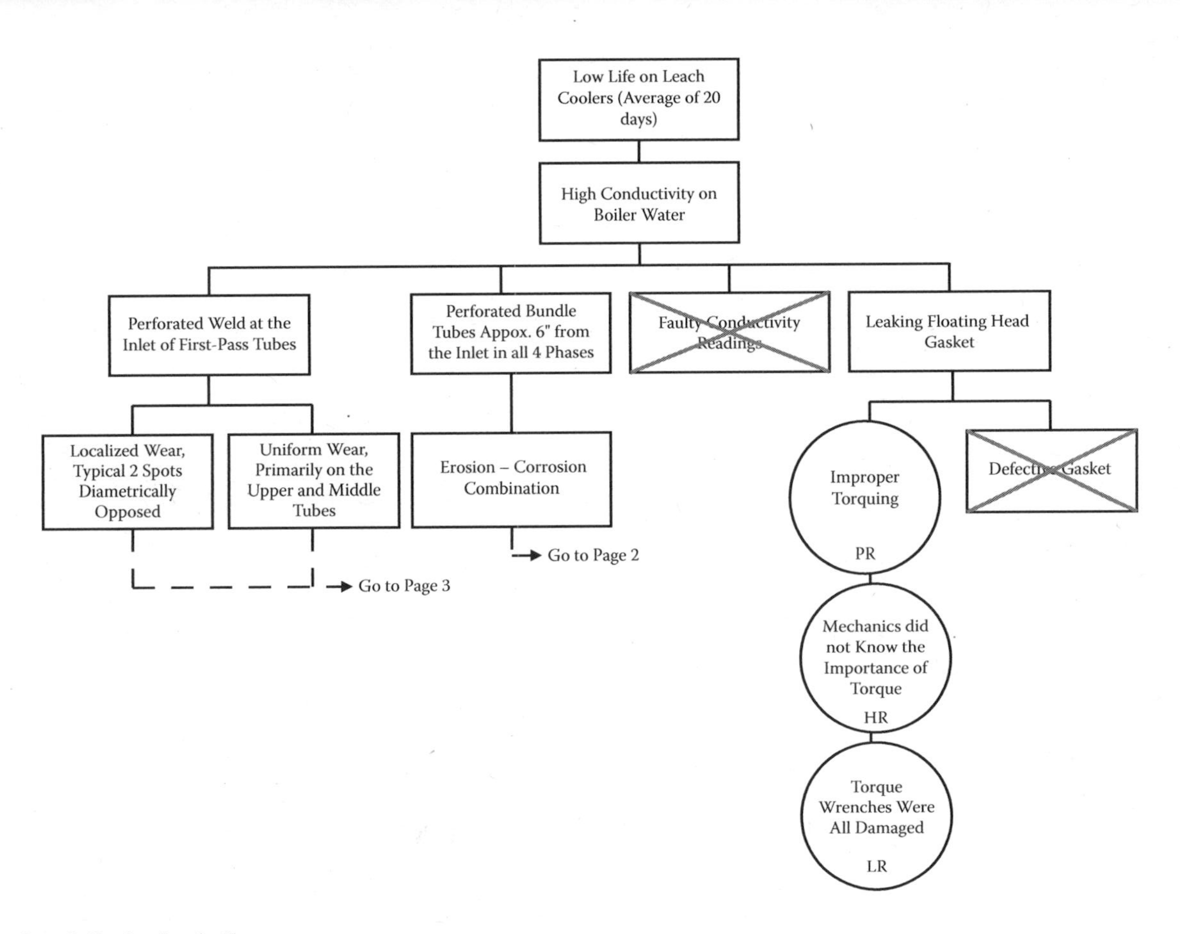

FIGURE 14.8 Leach Cooler Logic Tree.

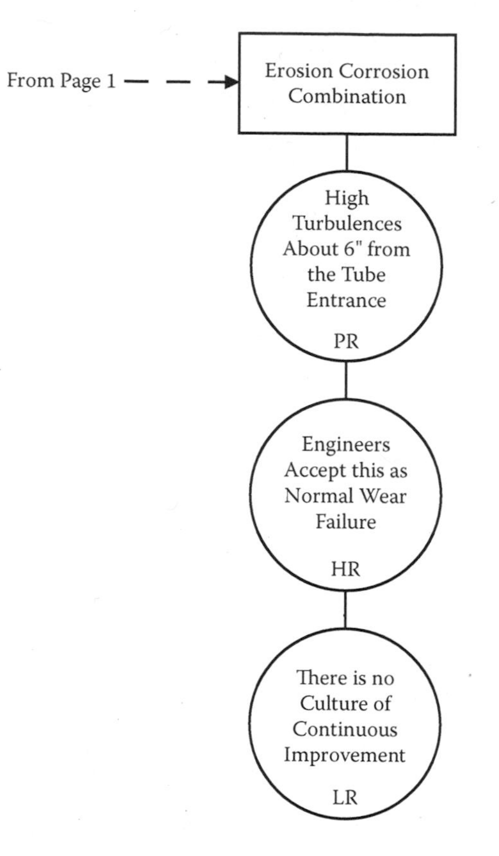

FIGURE 14.8 (continued).

have an explosive environment or for a fire to develop from the accumulation of these combustibles in the air chamber, grinding zone, classifier, burning lines, air inlet-duct, and/or the feeder whether the Pulverizer is in operation or in an idle state.

After extinguishing three pyrites hopper fires earlier in the day, there was an explosion in the Pulverizer that resulted in the tripping of the unit and activation of the inerting gas and fogging safeguards of the Pulverizer system. The plant was put into a safe operating environment and it was determined that a formal Root Cause Analysis, facilitated by the Reliability Center, Inc., be conducted on this important incident at this time.

Identified Physical Roots:

- Temporary Pulverizer operations less than adequate
- Upset Pulverizer operations less than adequate
- Calibration test inaccurate for existing Pulverizer design
- Coal quality issues

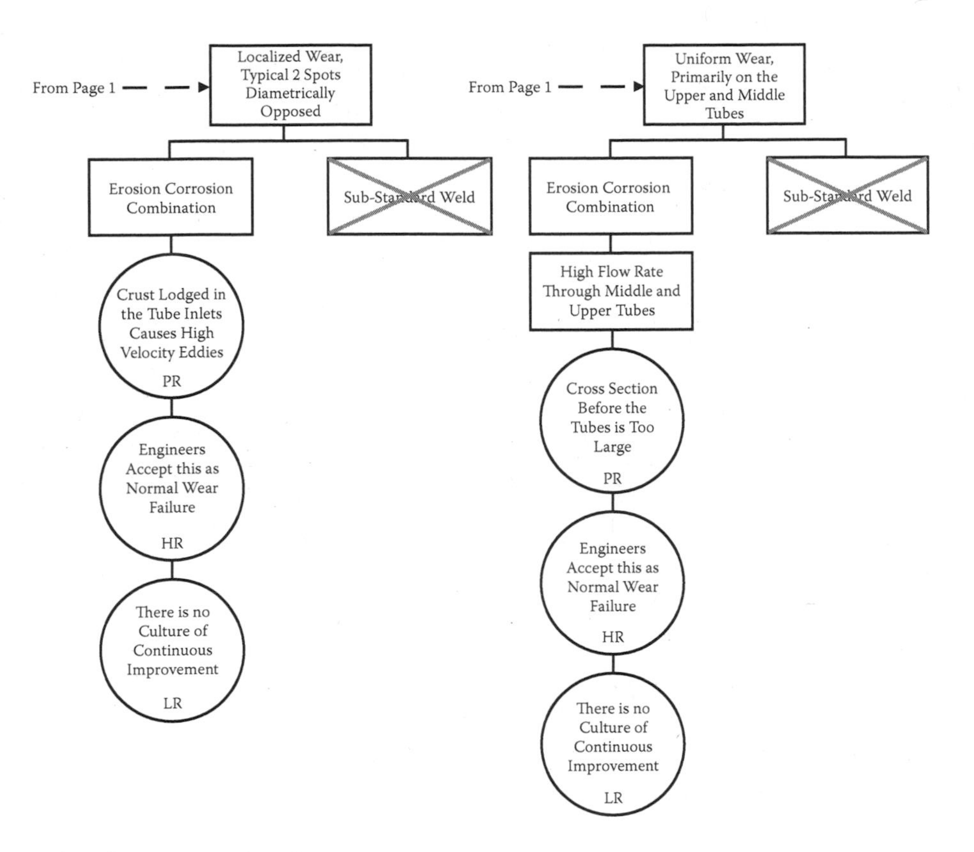

FIGURE 14.8 (continued).

Human Roots:

- Manual Pulverizer operations instead of automatic controls
- Inadequate management of change for new fuel (coal quality)
- Incorrect response to fires in the Pulverizer
- Design error—design locations inhibit proper calibration of Pulverizer
- Uncontrolled operator workarounds

Latent Roots:

- Counting on training to cover new fuel changes
- Locations of test ports are unique to this Pulverizer
- Operational procedures LTA (less than adequate)
- Project(s) is schedule driver (identified paradigm)
- Mega watts is primary (identified paradigm)
- Perceived time constraints (identified paradigm)
- Using standing orders in lieu of formal procedures

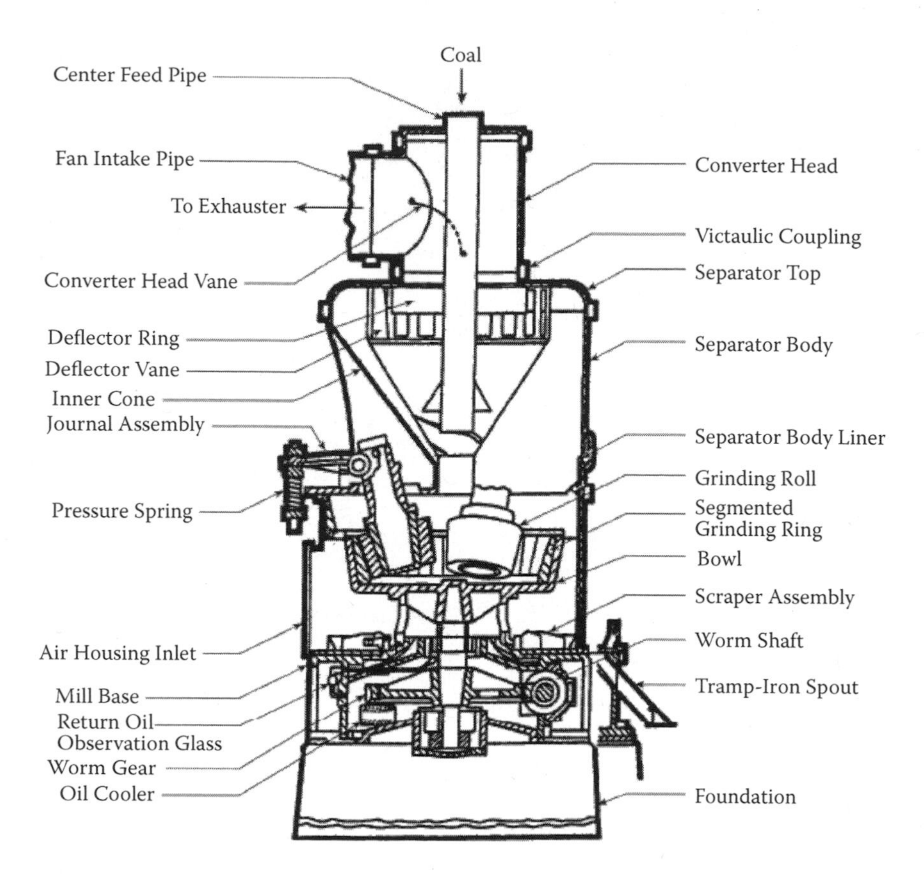

FIGURE 14.9 Typical coal Pulverizer.

- Procedures not released
- On-the-job (OJT) training less than adequate
- Formal Pulverizer training less than adequate
- Procedures for Pulverizer released in draft form

Implemented Corrective Actions:

- Rework Pulverizer Operational Procedures to bring them up to current requirements for safe and reliable operations of the equipment.
- Rework Standing Orders for upset operations to bring them up to the current requirements for safe and reliable operations of the equipment.
- Rework Calibration Procedures to meet the unique requirements of this particular Pulverizer and Hot Primary Air Duct.
- Address outage scheduling time constraints to include safe plant operations during formal and OJT training sessions.
- Update training to meet current plant operational requirements.

- Review OJT training for plant operations and update as necessary to meet current plant requirements.
- Update plant operational procedures to meet current plant requirements.
- Finalize formal procedures to replace the existing Standing Orders.
- Correct policy to allow procedures to be released in draft form.
- During formal training and OJT stress the importance of plant safety versus perceived time constraints.
- Educate plant personnel on the issues related to coal quality during training sessions.
- Educate plant personnel about safety concerns when dealing with projects versus plant scheduling priorities (safety always comes first).
- Review existing training to make sure it properly addresses new fuel issues.
- Rework Calibration Procedures/Test to cover the unique requirements of the Pulverizer.
- Rework the Instrumentation and the Calibration Procedures to cover the unique requirements of the Pulverizer.
- Correct air/fuel curve to achieve OEM's and NFPA standards.

Effect on Bottom Line:

- Elimination of all future Pulverizer explosions
- Safe and effective response to pyrites fires in Pulverizer(s)

Additional Comments:

- Consulting plus labor costs of company employees: approx. $100,000
- Findings of the analysis leveraged throughout entire corporation

RCA Team Statistics:

- Start Date: July/2007
- End Date: August/2007

Corrective Action Time Frame:

- The time frame from the time that the event occurred to the time the corrective actions were taken was 1 year.

Core Team Members:

- Joseph Brown (Principal Analyst)
- Ron Hughes (RCI RCA Mentor)
- John Laag
- Kenneth Kearney
- Richard Stewart
- Don Abati

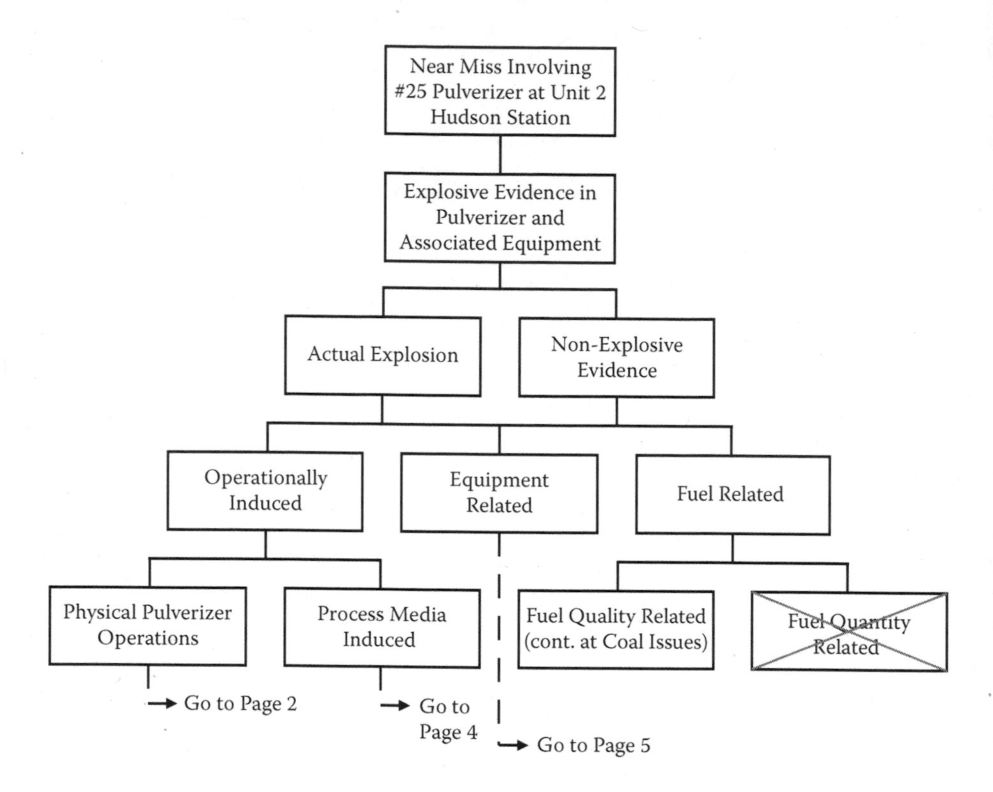

FIGURE 14.10 Pulverizer explosion Logic Tree.

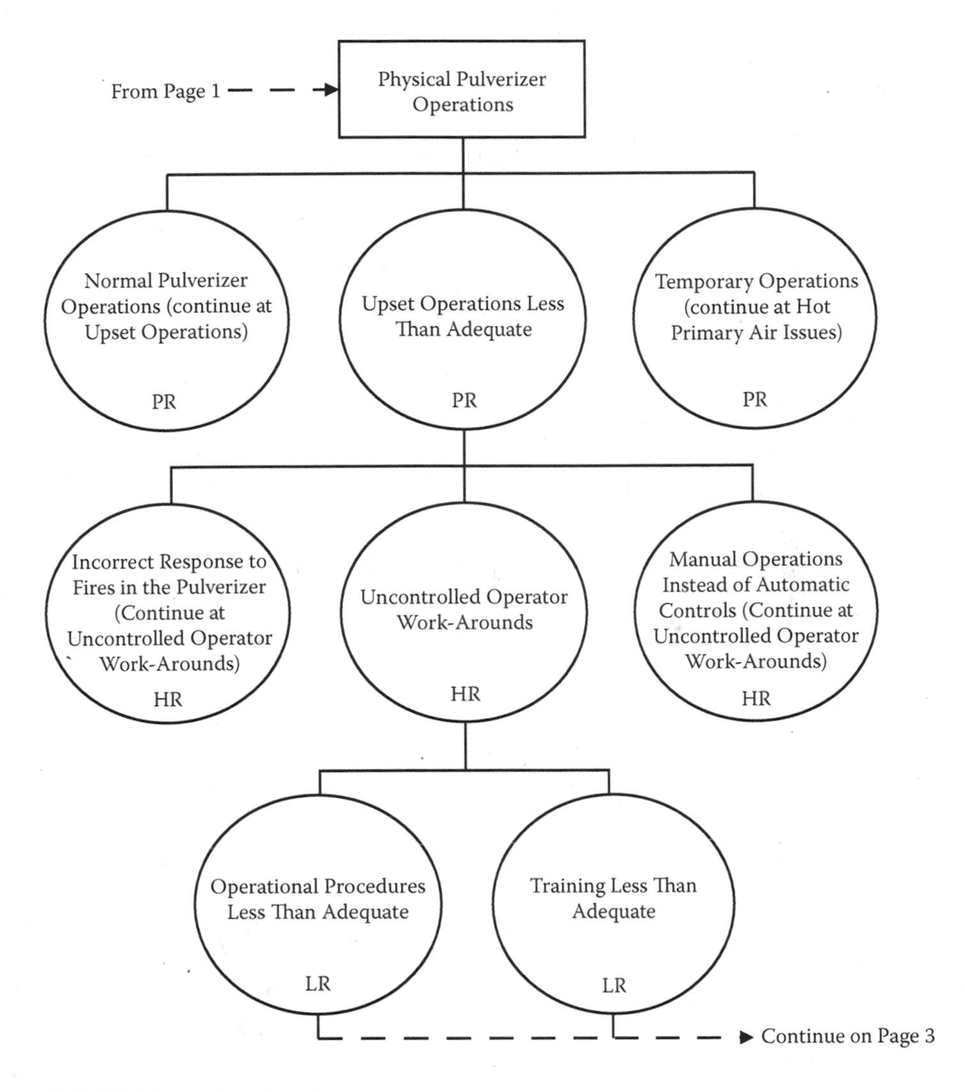

FIGURE 14.10 (continued).

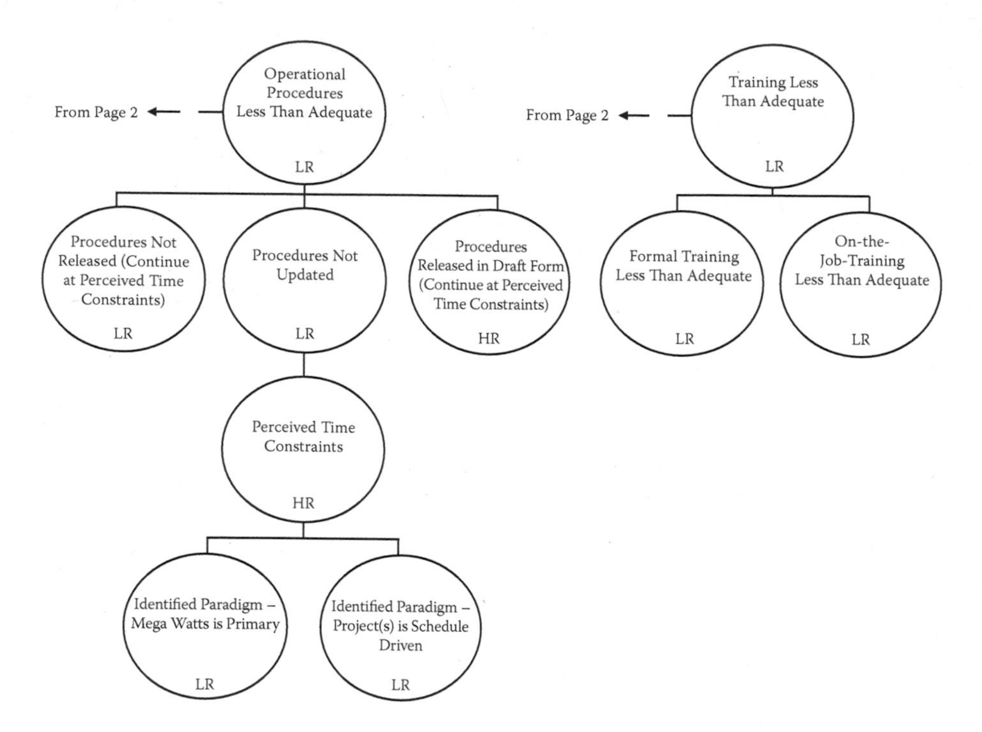

FIGURE 14.10 (continued).

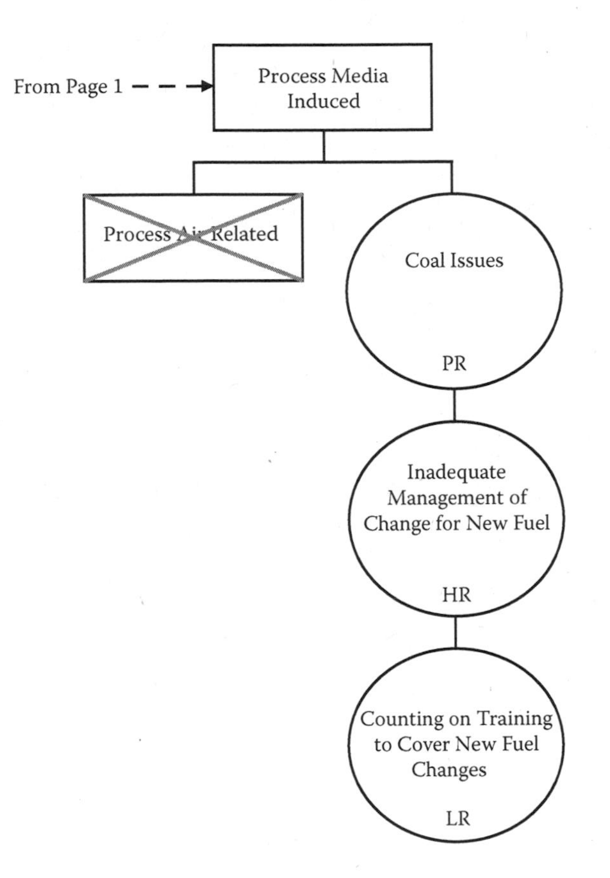

FIGURE 14.10 (continued).

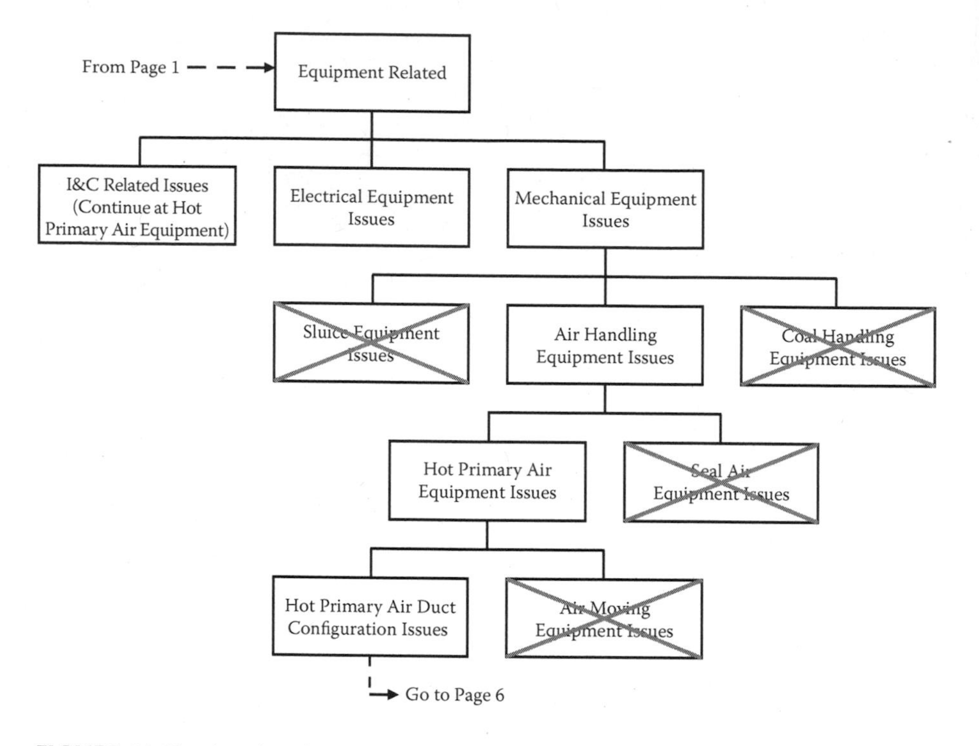

FIGURE 14.10 (continued).

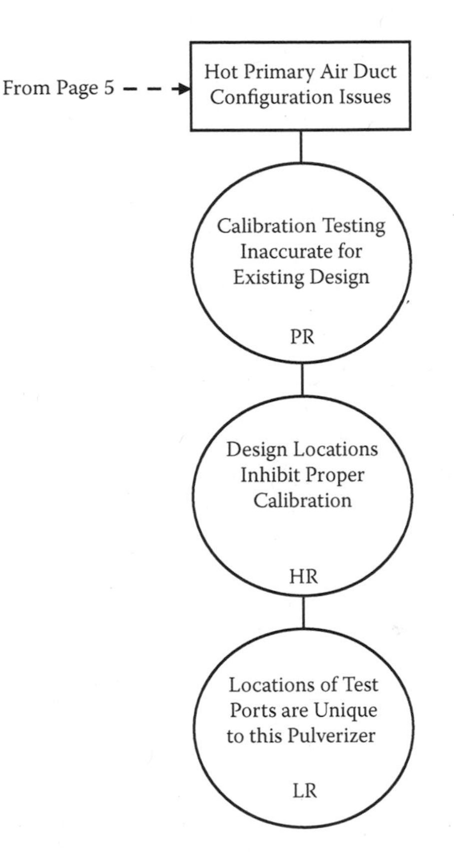

FIGURE 14.10 (continued).

Index

J

K

L

M

N

O

P